PLAY: AN INTERDISCIPLINARY SYNTHESIS

Edited by

F. F. McMahon
Donald E. Lytle
Brian Sutton-Smith

PLAY & CULTURE STUDIES, Volume 6

Co-published by arrangement with The Association for the Study of Play

Library of Congress Control Number: 2004113706
ISBN 0-7618-3042-1 (paperback : alk. ppr.)

Contents

Editor's Note

I am grateful to my co-editors, Brian Sutton-Smith and Don Lytle, who provided invaluable scholarly assistance for this volume. The interstitial introductions were written entirely by Brian Sutton-Smith. Don Lytle completed the final editing, indexing and formatting, and prepared the camera-ready copy necessary for the publication of this volume.

Felicia Faye McMahon
PARC, The Maxwell School
Syracuse University

ACKNOWLEDGMENTS

The editors of this volume on behalf of TASP wish to sincerely thank the many scholars who provided expert review of the 14 chapters in this book. We acknowledge your professional dedication within your respective disciplines to the scholarly study of play.

Felicia Faye McMahon
Donald E. Lytle
Brian Sutton-Smith

Introduction

The Association for the Study of Play (TASP) is an interdisciplinary group of several hundred international scholars who meet once a year to make known their latest research on play. This volume is the sixth in the present *Play and Culture Studies* series. It was preceded by one year of four issues of the journal of *Play Theory and Research* (1993) and prior to that, five years of the *Play and Culture* journal (1988-1992). The publishing efforts of this group, however, began in 1976 and proceeded with 10 variously titled volumes issued by the Leisure Press between 1974 and 1987. What impresses a reviewer is the great diversity of play accounts that are contained in these three decades of research records. The present volume with fourteen articles continues this record of diversity. In addition, however, editor, Brian Sutton-Smith of this volume has written another but smaller "interstitial book" in between sections I - V.

Like the efforts of scholars of earlier classic book length studies on play, notably, *Homo Ludens* (1949) by the historian Johan Huizinga; *Play, Dreams and Imitation* (1951) by the psychologist Jean Piaget; *Animal Play Behavior* (1981) by zoologist, Robert Fagen; *Dionysus Reborn* (1989) by classicist, Mihail Spariosu; and perhaps *The Ambiguity of Play* (1997) by educator-folklorist Brian Sutton-Smith, the editors of this volume attempt to make unified sense of the great diversity in play scholarship. The aforementioned classics in play books as well as thousands of other smaller scale publications on play have in various ways supported the practical importance of play in the lives of all of us, animal and human. Yet because of the continuing complexity and diversity of these studies there is no consensus on what play is and what it does, even though it is being given an ever increasing entertainment centrality in the modern world (Gabler, 1998).

The following volume is organized into four major sections: First (Part I), on the basic data about the evolution of play, second (Part II), then proceeds to evidence about universality from cross cultural comparisons, third (Part III), provides samples of the role of play in school socialization, and fourth (Part IV), concludes with some of the varying types of leisure as play to be found among adults. The references for this introduction and the interstitial material appear as a final bibliography. The other authors' references appear within each of their respective chapters.

Part I

Evolutionary Resonances for Play

CHAPTER 1, ROBERT FAGEN: PLAY, FIVE GATES TO THE EVOLUTION AND PATHS TO ART

We are greatly privileged to have this section introduced by Robert Fagen, the author of the all time greatest research review on *Animal Play Behavior* (1981). In his present article, he focuses with great scholarly depth on the origins of aesthetic behavior in organisms even prior to the origins of play. Our own approach for this section, however, is to consider evolution in terms of the early mammal history of play and to introduce his contribution within that framework. The editors need to stress in doing this that we do not mean to take attention away from his remarkable and highly original formulation of the evolutionary emergence of aesthetics in living creatures.

Most evidence suggests that the relevance of play to evolution arrives in prehistory with the appearance of warm-blooded mammals about 65 million years ago. The issue for us is to name the processes within which play first enters this mammalian situation. We suspect that there are at least six processes which may be relevant for understanding the genesis of play.

The First Process

The first process, described by the late Stephen Jay Gould (1996) is that evolution emerges as essentially a font of variation on all levels, genetic and behavioral. He says that without such variation natural selection would be meaningless. Variation is the essence of evolution and this applies to all kinds of functions. Further, he suggests that these emerging variations can be described as inherently redundant, flexible and quirky. A play theorist can easily savor these three concepts as potentially descriptive of various kinds of play, although because they describe all kinds of evolutionary variation. They would not be specific to play alone.

The Second Process

The second process described by Robert Fagen here is that all living creatures show preferences for some, but not others of the Gould-like variations available to them in sensations, perceptions, experiences and ideas. They prefer these selected variations despite their irrelevance to adaptation. All of which means that on two levels, random variation (Gould) and non-adaptive preferences (Fagen) there are already labile ways of functioning which hint at the nature of what play might come to be like. Fagen says, indeed, that the scope of these non adaptive preferences seem "so vast and all embracing that they furnish a convincing biological substrate for the origins of animal play itself." That does not, of course, mean that they constitute the complete meaning of play anymore than Gould's random variations constitute the complete meaning of play. But they do at least suggest that variability in multiple ways in native to all creatures and is an inherent possibility in all kinds of functioning.

The Third Process

The third process is outlined by play expert Gordon Burghardt (1999; see also Fagen, 1995) that the warm blooded animals and birds that emerged 65 million years ago, because of their immediate immaturity, required considerable protection from the normal predatory dangers of everyday life. But as such protection decreased the stimulation available to them their need also for big-brained extra stimulation may well have been gained through the parents instigating play with their offspring. This importance of the protective basis for the play responses is refreshed by more recent findings. Some pre animal reptilian creatures such as tortoises and giant lizards, which have not been observed to play in their usual dangerous environments, do appear to produce novel play like behavior in their non-dangerous zoo confinements (Burghardt, 1998). While this is a reasonable suggestion about a necessary or useful play context, it is still not a full explanation of how the above random variation, specialized preferences or parental stimulation turn into play itself.

The Fourth Process

The fourth process will be described here in two forms. The first is by Fagen in the present article and the second from our own work on the centrality of emotions in most forms of play (Sutton-Smith, 2001a & b, 2003). Thus, Fagen describes play as emerging primarily as a system for freeing the organism from the underlying biases (mental, affective or physical) in its system of non-adaptive preferences. In doing this, he says play becomes self organized, has a novel character and is intrinsically rewarding (as in the second process above) and can deal with unpredictable emergencies as well as develop a taste for them in its own sphere.

Sutton-Smith, on the other hand, places the predatory character of the mammal life process in the third process described above as the critical context for the emergence of play. Thus, the typical life of all living creatures requires incessant exploratory activity both to avoid dangers and to gain food (Berlyne, 1960). Exploratory activity is normally repeated until habituation occurs and is normally motivated by involuntary emotions such as those of anger, fear, disgust, shock, and so forth. Current research indicates that underlying emotions are always accompaniments of performances activities of any sort (Damasio, 1994). What happens in play then is that the affectively based exploratory repetition when it reaches habituation, on some occasions, can become transformed into play. There is indeed a considerable literature which illustrates this occasional transformation from exploratory to play behavior (Hutt, 1971). The position taken here and by others is that play when it occurs converts affective distress into relief (Shore, 1994). There is some evidence of infants themselves converting their "serious" activities like sucking into carefree playing at sucking with smiles on their faces (Piaget, 1951). And the teasing data show that parents can induce apprehension in the baby by making loud sounds or funny faces etc. but they also laugh and smile while doing it, so that the baby learns that these exaggerations are not harmful and, in due course, initiates these same kinds of errant behaviors themselves (Bruner, Jolly & Sylva, 1976; Briggs, 1998; DeLoache & Gottlieb, 2000). We have to assume that such transformations are a kind of variability, which is easily triggered at habituated points of responses to affective danger when they are given over to symbolizations of relief. It is notable that such "relief" is characterized by performances which are exaggerated, grotesque, nonsensical, and so on. While these performances can be original, they are also are typically very repetitive and "quirky" as they make less than usual sense. This is very important because this exaggerative folly and fun that is central to play is a guarantor that these virtual play behaviors will subsume the more dangerous real affective behaviors from which they emerge. Elsewhere play is defined as a parody and ironic because it both represents the affective metaphors of danger and yet defies them inversely by contrary masteries while paradoxically also simulating their real life affective unpredictability by the play's open-ended outcomes (Hutcheon, 1985). Some illustrative play examples are:

> **Contests** which represent anger but inverts it through tactical and strategic masteries;

> **Festivals** which represent bonding but invert it through selective communal identifications;

> **Gambling** which represents risk but inverts it through fateful courage;

> **Hazing and teasing, which** represent shock but invert it through acquired social resilience;

> **Trickster play** which represents disgust but inverts it though iconoclastic imagination;

> **Cynosural play** which represents pride but inverts it through histrionic charisma.

The Fifth Process

We must ask now what if any are the biologically or culturally adaptive values of such playful inversions of emotions. For example, these few examples are sufficient to assert that play is about the modulation of involuntary by voluntary emotions. It replaces the involuntary emotions such as anger and fear with the voluntary emotions illustrated by exuberance and playfulness. In these terms, play is about converting stress into enjoyable but ridiculous excess. Fagen perhaps confirms such a process when he says, "social play is imparting emotional intelligence to agonistic behavioral domains" (p. 35). Play modulates blind fear and blind trust and produces more flexible forms of predatory and predator avoidance behavior" (p. 34).

Whether play can be both such a modulator of fear etc. and as well be a source of the creative spirit, as he defines it, is open to further interpretation. It is not improbable that Fagen's central interest in art has made him disposed to exaggerate the amount or kind of creativity to be found in play. One is after all not generally much impressed by the creativity of play even if still excited by the massive and redundant amounts of excitement in sports, carnivals, gambling, and foolishness as presented above. The function of play seems to be harnessed more fundamentally to the modulation of our rather primitive emotions than to be causative of any higher levels of artistic creativity. Play seems to illustrate most clearly that we can control those emotions while still enjoying them in these constrained forms. This does not mean one cannot be playful in the arts, but that maybe it's more about the virtues of the free emotional spirit that can be adopted, rather than about the qualities of creativity that might be available from play itself. Perhaps this fifth process has most to do adaptively with the way in which these affect modulation processes have random or other transfer effects into alternate systems of everyday reality. The transfer effects are those that have attracted the most attention in modern socialization (psychological and sociological) research. It would be assumed that the major socialization affect from play as defined in process four would be some form of affect regulation. It might increase the management of stressful emotional controls both in and out of play which is what Fagen suggests and some animal research supports. This has also certainly been a major assumption within the various schools of psychodynamic play therapy during the past century. There is also some slight quantifiable developmental psychology evidence of that character (Carson, Burks & Parke, 1993).

The Sixth Process

The sixth process to be presented here is a more complete reification of Fagen's view that play is intrinsically motivated, and that it need not have to be adaptive but can lead to multiple biological consequences despite its own virtuality. For example, play, perhaps even initially, but certainly finally, becomes not just a way of simulating the emotional-survival conflicts in the external world, but rather is also a way of substituting the play itself as a more preferable kind of alternative emotional reality. In these terms, play can be said to be an escape from the world rather than a reconciliation to it (Tuan, 1998). It is important to realize that the concept of adaptation which is typically taken as the supremely important reality within evolutionary theory, by contrast includes in every day life a host of "adaptations" which themselves can be inane, boring, humiliating, chaotic and finally mortal. The paradox is that play which is supposed to assist orthodox adaptation into that partly indifferent world of adaptation can also become instead a replacement for it, and as such becomes for many a superior form of alternative adaptation. Play need have nothing to do with typical survival oriented adaptation, and yet because it provides such a good time to its participants, it leaves them often happier about their usual life circumstances. So play may be said to have become adaptive by ignoring the usual norms for such adaptation. This is, of course, just another possibility that arises from having such a good time.

CHAPTER 2. PEGGY O'NEILL-WAGNER: VIDEOTAPE ENTERTAINMENT MAY FACILITATE RECOVERY FOR MONKEYS IN A CLINICAL SETTING

The discovery that the DNA difference between us and various classes of primates is extremely small, a few percentage points in most cases, has given us all a much more avid appetite for the parallels between us. There is in fact a whole new literature focusing on our resemblances, which are much greater than we had once supposed (Diamond, 1992; Wrangham & Peterson, 1996; etc.). The present brief anecdotal piece is a part of this modern evolutionary exploration although what we seem to be seeing in these present monkeys is leisurely diversion rather than active play. The monkeys seem not unlike human television audiences, in sometimes playing along parallel to the shows, or in being annoyed at interruptions, or in compensating for life's other failures, or in forgetting their own illnesses, and finally in making sexual connections. But they can go even further than this. An article in Discover magazine [May 2002] included a description of the play behaviors of chimpanzees used in aids research and are now destined to live out their lives in quarantine. In spite of the rather unpleasant experience of being injected with various incurable diseases, the chimpanzees engaged in play, not only, among themselves but went so far as to engage in teasing their human captors. Some of these teasing behaviors included spitting water at the human researchers and then laughing at them. This kind of play behavior remained consistent throughout the chimps' captivity. Not only did they delight in teasing humans as well as each other but the human caretakers noted that the chimps cleverly varied the patterns of water spitting to confuse the researchers, thus effectively increasing their play through trickery.

What this data does is to reaffirm the point implicit in the prior chapter that play is a long time animal evolutionary response and theories about it need to keep this in mind. This remark is leveled at the modern tendency to attempt to confine play to some more abstract and creative human forms as infinite but not finite, or as flow which is not self consciousness, or as amateur and not professional when play at base is a kind of fluctuating emotional and performance excitement enjoyed by all mammals (Lindquist & Handelman, 2001).

CHAPTER 3. THOMAS L. REED: A QUALITATIVE APPROACH TO BOYS' ROUGH AND TUMBE PLAY

As Thomas Power (2000) says in his masterly research review, *Play and Exploration in Children and Animals,*

> Although it is impossible to prove that any behavior pattern in humans has an evolutionary basis, the research on play fighting provides the strongest empirical support for this notion in the area of children's play. As one would expect for a human behavior pattern with a long evolutionary history, play fighting is found cross-cultures and shows structural similarities with behavior of our animal relatives as a function of our phylogenetic proximity to them" (p.188).

The present persuasive chapter is relatively exceptional in this area though there are earlier reports in TASP about the kindness of the boy playfighters with each other (Wegener-Sphoring, 1989). It is clear that the emphasis in most such play fighting is upon a mask of "toughness" behind which there is nevertheless an often unspoken caring. If we assume that there is a relevance of play fighting to warfare then its seems likely that it has to do with creating a readiness for the kind of intragroup trust that is useful when it comes to fighting as teams. There are world wide analyses of male initiations and hazing which also suggest that the shock and harassment that the initiates undergo makes little sense except as a way of forming a common bond amongst those who have been so embattled (Bayer, 2001). The inhibition of verbalized feelings about emotions is often central to male group socialization. But then so is the containment of any real stress in these play fighting games. Though the games contain the mock emotions and mock gestures of warfare and aggression, they are in fact play forms in which the performances have much to do with avoidance of emotional extremity though under the cloak of representing the same. The plays have as much to do with controlling the emotion (anger) as they have with representing it through some indirect attack formula such as in the present case "queer" or chess etc. The control of stressful emotion in these forms of play which is widespread in humans and animals makes a strong claim once more for the containment of "emotionality" to be a critical component of any adequate play theory. Interestingly, this may be the underlying thesis in Mariah Burton Nelson's conjecture that adult males must find an avenue to express their affection as well as their anger at women, in socially acceptable male sports such as wrestling and football. However, there is no data to support Nelson's claim that notably absent from the gay male sports culture is football and that gay men do not 'need' football because they have "created alternative culture in which they can celebrate masculinity, male bodies male sexuality" (1994, p. 120). The evidence seems to be to the contrary. Nelson also suggests that some same-sex play activities that exclude girls 'cements' the boys' place in the dominant class and that is obviously one in which emotions are masked. We might sum up Reed's chapter by saying somewhat poetically that in play fighting it is possible for play as a gesture of courage to be a mask for play as a gesture of affection.

Chapter 1

Play, Five Evolutionary Gates, and Paths to Art

Robert Fagen

OVERVIEW

Animals, including fish, frogs, and birds, may show preferences of unknown origin for certain colors, shapes, sounds, or other stimuli, sensations, perceptions, experiences, and ideas.

The consequences of actions based on preexisting preferences do not contribute directly to the animals' personal survival, or to survival of other animals carrying any hypothetical genes on which specific or general preferences may be based (e.g. in close genetic relatives). These preexisting preferences may be said to have no economic or eugenic significance, whether immediate or delayed. But if A's preexisting preferences for B's traits contribute to B's reproduction, and even to B's survival, then preexisting preferences become a factor in evolution, comparable to artificial selection.

Experiences that have rewarding qualities may in addition accompany expression of a preexisting preference. These preferences would then be said to exist solely for the sake of the experience that results from expressing them.

When humans pay attention to something, and they do so with disinterested motives, we say that they are "contemplating." This contemplative stance defines the domain of aesthetics. Preferences for objects of contemplation, and the contemplative act or stance itself, do not (by definition) contribute to personal survival, because the motives involved are purely disinterested.

Biologists use the term "evolutionary aesthetics" confusingly. Sometimes they mean the tendency to prefer, and to respond emotionally in particular ways, to features of the living and nonliving environment whose selection by the subject will contribute to that subject's inclusive fitness and/or reproductive success. At other times, they restrict the term to contemplation and use "aesthetic" as a catchy synonym for "nonadaptive" or "preexisting nonadaptive."

Preexisting sensory biases are important for evolution in theory and in fact. In particular, sensory preferences of unknown origin form the basis for a dimension of evolution distinct from that occupied by preferences for environmental features that enhance the subject's survival. Computer models of animal sensory preferences can behave analogously to real animals. Therefore, both real animals and computer models can evolve sensory preferences that do not increase reproductive success by any conventional means and may even decrease it.

Some treatments of evolutionary aesthetics view nonadaptive sensory preferences as a possible biological basis for an evolutionary theory of aesthetics and even for an evolutionary psychology of human art and literature. But most discussions still define aesthetics in terms of preferences for features that enhance the subject's own survival, or at best lead to selection of a mate whose characteristics will benefit the survival of the subject's hypothetical genes for the preference.

Thus, they seek to explain human art and literature in terms of economics and/or eugenics.

Discovery of nonadaptive sensory preferences has extended the scope of evolutionary models far beyond their origins in theories of evolution by natural selection. This discovery has also generalized Darwin's concept of sexual selection by extending the scope of conventional sexual selection models. That sensory preferences can be nonadaptive, yet evolve means that scholars must reevaluate previously-hypothesized roles for animal play in the evolutionary origins of art. The best evidence for nonadaptive sensory preferences comes from bird and fish species not known to play at all. There is no reason to suppose that such preferences should not also occur in mammals, including humans.

Darwin originally considered preferences of unknown origin in the context of sexual selection. Mate choice by females was Darwin's specific concern. Since Darwin, the concept of preexisting preferences has further expanded to include food, habitat, and similar ecological characteristics.

The implications of preexisting preferences have also expanded to include long-term developmental as well as short-term ecological effects. For example, a preexisting preference could produce experiences with enduring effects on modifiable developmental templates.

Because light and sound, and vision and hearing, lend themselves best to experimentation, the literature of nonadaptive sensory preferences emphasizes these two modalities. But other senses, including smell, touch, taste, kinesthetic, and vestibular must be involved as well. Indeed, nonadaptive sensory preferences are only one kind of nonadaptive preference, for higher mental processes are interactive and synthetic, linking sensation, perception, emotion, and cognition, and defining a realm of experience that includes both self and environment. It seems obvious that animals can also evolve nonadaptive preferences for patterns of touch and movement, for patterns of emotion and cognition, and for qualities of experience. Indeed, the scope of these preferences seems so vast and all-embracing that they furnish a convincing biological substrate for the origin of animal play itself.

The biological processes of sensation, perception, emotion and cognition have emergent properties that enable them to transcend adaptation. In turn, the biological processes that constitute play have their own emergent properties. The most important of these seems to be that play can free the underlying processes from their biases, just as bias freed the lower-lying motor effector processes from their hypothesized primitive dependence on the dictates of reproductive success. Play may even have been selected as a way to unbias the mind, the emotions, and the senses and thereby to counteract the extreme effects of nonadaptive preferences. Indeed, one of the most pervasive and experimentally well-demonstrated effects of animal play is its ability to impart flexibility and adaptability to an otherwise-rigid biological machine. As such, play seems well suited as a source of the creative impulse that provides the spark for art. By recognizing nonadaptive preferences and by differentiating them as such from play itself, we may be able to make substantial progress towards an understanding of art in humans and towards a clearer vision of human nature.

So far as is known, classical Indian philosophy first developed a world-view of play in evolution. Cosmologies in which creation is an ongoing process, represented by and situated in play, can be difficult for Western minds to grasp. These cosmologies may appear particularly opaque to those trained in scientific modes of thought. But such cosmologies are inevitable consequences of this essay's scientific analyses of art and play.

INTRODUCTION

Are play and art somehow connected? This chapter argues that the creative spirit common to human play and human art has nonhuman origins in the play of animals. But the human sense of beauty essential to art may have little to do with animal play. The chapter further argues that the aesthetic sense also has nonhuman origins, albeit distinct from play. The human sense of beauty has biological roots in arbitrary sensorimotor and cognitive biases that biologists find in many species of animals.

This chapter is not strictly linear. For greater ease in reading, please keep the essay's conclusions in mind (and please see Conclusions, now). Thinking about art and play can prove extraordinarily productive. Ultimately, it opens new gateways to that most difficult and important of all ideas, the classical Indian conception of the ongoing play of creation, so difficult for Western-trained minds to grasp, and so eloquently and learnedly presented (early on, to TASP audiences) by Don Handelman (Handelman 1992; Handelman & Shulman, 1997).

Play and Culture Studies readers will understand that not all biological perspectives are identical. The humanities and social sciences have survived repeated challenges from biological determinists, including some who base putative syntheses of all human knowledge on narrow biological views of human nature. This chapter approaches aesthetics from biological and evolutionary perspectives in which beauty, play, and intelligent order are fundamental. Its arguments rest on the facts and theory of evolution applied to a universe which is demonstrably neither fair not just. Readers will learn that not every biologist believes that genes rule, whether directly or indirectly.

The claims of sociobiology and evolutionary psychology are hard to ignore. To do so would be about as easy as ignoring a live-in adolescent at close range. These postmodern biologists, and their counterparts in the humanities, have declared victory in one intellectual battle after another, rendering the projects of classical aesthetic philosophy increasingly problematic, and the very experience of beauty antique. With the phrase "It's in our genes" on everyone's lips (though few people seem to ask just how it got there), evolutionary, indeed sociobiological approaches to problems of aesthetics and of poetics, and even to beauty itself, are very much the order of the day (see, e.g., Feist, 2001).

I am a biologist, but not a "genes" person. This chapter responds to recent biological analyses of aesthetics and seeks to introduce relevant findings to an audience of non-biologists. Lumsden (2001) correctly points out the inadequacy of textbook genetic determinism for studying the human mind. Interactions govern this problem and require new domains of discourse. Different language, one specifically designed to address interaction using domain-specific grammar and syntax, is required (Lewontin, 1974). Efforts to achieve this goal virtually define the history of biological thought on genetics and development, from epigenetic landscapes (Waddington, 1957) to metaphoric walls constructed from bricks and mortar. Similar trends are evident in the humanities. Challenges that critical theory and postmodernism once posed both to humanists and to biologists helped spur resurgent scholarly interest in beauty and in aesthetics, a rich and readable body of humanistic studies on these topics (e.g. Brand, 2000; Clark, 2000; Elliott, Caton, & Rhyne, 2001; Fisher, 1999; Hampl, 1999; Scarry, 1999), and a splash of interesting websites (e.g. absolutely, 2000). Intended in part to counter biological sallies such as those by Orians (1996), Symons (1995), Thornhill (1998), and Wilson (1998), these newer studies (Lambert, 1999) redirected previous approaches.

Poetics, aesthetics, and biology now seek a common language. Cognitive science has offered to help (Richardson, 2001). The problem requires fluent handling of qualities (Goodwin, 1994; Fielding & Lee, 1991), metaphor (Lakoff & Johnson, 1999), and narrative. Separate domain-specific procedures for verbal and for non-verbal information might be necessary, but the success of object-oriented languages in computer science may offer hints of common ground.

Biologists seeking to reformulate aesthetics and poetics in evolutionary terms have not yet confronted the latest work in the humanities. Moreover, the work of leading aestheticians like John Dewey (1934), Francis Sparshott (1982, 1995, 1998) and Yi-Fu Tuan (1974, 1993) is too often neglected. In this essay, I seek to juxtapose contemporary evolutionary biology with relevant and cognate work in aesthetics, poetics, and cognitive theory. I chose to limit the scope of this chapter and its bibliographic coverage to those sources that I found most useful in tracing the history and findings of Darwinist aesthetics. I had to omit fascinating material: the behavior and evolution of bowerbirds, believed by some biologists to represent an evolutionary pinnacle of the aesthetic sense in nonhumans (Diamond, 1986; Johnsgard, 1994); the discovery that common chimpanzees at Gombe Stream find falling water so sublime that they sometimes go on long hikes in order to gaze at waterfalls, and perhaps to contemplate as well (Bauer, 1976); and the long history of ethological speculation about the origins of human art (e.g., Morris, 1962). This chapter serves only as a preliminary sketch and as a bridge to further work.

Factors of Evolution: Everybody Into the Gene Pool

Stephen Mo Hanan's play *Everybody Into the Gene Pool* takes an irreverent, humorous and hugely entertaining look at birth, death, and creation. His work wisely reminds us that there can be more than one story about human origins and that evolution itself is many-gated. Michael Ghiselin (1969, 1974) reminded biologists of the historical importance and rich implications of classical economics in Darwinist thought. In addition to these economic factors that frame the intellectual heartland of Darwinism, eugenical factors (Ghiselin, 1974, pgs. 148, 181, 182) may be invoked to explain the history of life, especially by those who advocate gene-level approaches to evolution. Darwin himself identified, and Ghiselin further discussed, a third kind of evolutionary factor: the aesthetic (Ghiselin, 1974, p. 256, 262, 263). Aesthetic factors are based on individual preferences of unknown origin, as exhibited, for example, in female choice (theory of sexual selection, Ghiselin, 1974 p. 130-135, 175-186) and in domestication and social evolution (theory of artificial selection) (Ghiselin, 1974, pgs. 130, 191, 258, 262).

Schmalhausen (1949) also spoke of "factors" in evolution, with emphasis on stabilizing selection, but in a different context from that considered here, though with a common emphasis on developmental mechanisms as the key to understanding pluralistic evolutionary modes.

Biological Determinism and Determined Biologists

But can biology deliver what its advocates claim? Biological analyses of human behavior and culture are dogma in some circles, anathema in others. It's not new for a particular approach to generate controversy, and to politicize an entire subject for a generation or more. An approach to art via evolutionary aesthetics presumes that one royal road, or several less-trodden ones, can yield sufficient novel insights to constitute a substantive contribution to understanding. The strong form of this approach is to embrace a particular dogma as the sole and entire truth, to proceed to view everything in its light, and to ignore everything that does not reflect its light. This approach, though it hardly honors the human mind, offers advantages. Rather than getting bogged down in aesthetic philosophy or in the ethnography of human art, biologists have a program and a way to accomplish something. Thus, when we look over the field, we see systematic and concerted bodies of knowledge termed, for example, Freudian aesthetics, Marxist aesthetics, Darwinist aesthetics, feminist aesthetics, or masculinist aesthetics. The advantage of powerful, rich belief systems like these is that they can and do come up with explanations for many things, and that they can very gracefully and inconspicuously usher out everything that doesn't fit. The disadvantage is that they have circumscribed domains and finite lifetimes. Heroics, however Promethean in scope, can accomplish little in a post-heroic age when fundamental ideals of scholarship must don clown makeup to survive and where heroics are best left to comic-book characters like Superman, Robin, Lynx, and Zoz (see Anderson 2001).

PROLEGOMENATOAN EVOLUTIONARY AESTHETICS

Aesthetic Factors in Contemporary Darwinism

Evolutionary aesthetics seeks to apply insights and modes of thought drawn from the biological sciences to aesthetic philosophy, to literary theory and practice, and to the ethnography of art. For their justification, these studies appeal to contemporary evolutionary theory, with Darwin's gradualist view of continuity between human and nonhuman species furnishing the necessary historical framework. Often, studies of this kind draw on current modes of thought current in cognitive science. Though this new field of biological inquiry has not yet achieved conceptual unity, its polymorphism is appealing, and I fervently hope that no latter-day New Synthesis lurks in the wings.

A survey of the field identifies several large, widely-separated sheaves of studies, each so isolated from all of the others that a bibliography containing each one's current works and the references cited therein would show almost no overlap with any of the other bibliographies. Since each sheaf represents the work, completed and in progress, of large numbers of prolific scholars, the sheer volume of contributions is staggering. This environment selects strongly for user-friendly packaging. An edited collection of papers or citations can give a symbolic name and concrete, take-home form to particular sheaves. But most of this material is still so scattered that no single existing compilation or review is adequate to summarize the field as a whole.

General overview

This review will be concerned chiefly with the work of biologists who followed Darwin in equating sensory biases of unknown origin with a nascent sense of beauty. But no discussion of evolutionary aesthetics would be complete without mentioning some other kinds of studies as well, especially since their makers also use the term "evolutionary" to describe their work.

Richardson (2001) compiled and annotated a bibliography on literature, cognition and the brain and included a number of works whose intent might be considered "evolutionary poetics." Advances in cognitive science and philosophy of mind seem increasingly linked to biological, indeed evolutionary ideas. Though Richardson's bibliography primarily covers poetics and literary theory rather than aesthetics and the visual arts, it indicates that the biologically-based insights of Damasio, Edelman, Gardner, Johnson, Lakoff, Turner and others have deeply influenced areas not distant from the theory and practice of aesthetics. In fact, an entire issue of the *Bulletin of Psychology and the Arts* (Feist, 2001) devoted to evolution, creativity, and aesthetics addresses many of the same themes from explicitly evolutionary perspectives. Although some authors in this collection seem overly fond of citing their own work (39 self-citations in one case, 27 in a second), other bibliographies in the volume are more balanced. In particular, that of Lumsden (2001) includes a representative sample of contemporary biological studies on the creative mind.

Preexisting sensory bias

Darwin's own ideas about a sense of beauty in nonhumans are the key to understanding a body of work that is rigorously evolutionary and offers a metaphoric relationship to the study of aesthetics and art. This set of studies involves nonhuman species, particularly fishes and birds, and is most closely associated with Darwin through his theory of sexual selection. Darwin's fundamental observation was that arbitrary preferences of unknown origin seemed to govern mate choice. As his discussion of female choice indicates, these preexisting preferences were primary in Darwin's thinking. In Darwin's notebook entries (Darwin, 1987) and published writings (Darwin, 1874, p. 326-331) on the topic, the special cases of "good genes" and "good mates," often mistakenly equated today with the whole of sexual selection theory, seem almost trivial by comparison to the basic biological fact of sheer, arbitrary preference—for Darwin, the prime mover of sexual selection. In fact, Darwin spoke of animals' "taste for the beautiful" in terms of a beauty represented by "any brilliant, glittering, or curious object" (Darwin, 1874, p. 326) and by "the most ornamented males, or those which are the best songsters or play the best antics" (Darwin, 1874, p. 330). Only after making these points did Darwin introduce the important topics of "vigor and liveliness" into his discussion (Darwin, 1874, p. 330). The theme of arbitrary preferences recurs in Darwin's discussion of birds' "taste for the beautiful" (Darwin, 1874, p. 626) and in his premise that animals would choose mates "characterised in some peculiar manner" (Darwin, 1874 p. 941).

The great British ecologist Charles Elton (1930, 1935) also realized that animals sometimes make choices based on factors that are not exclusive physical. In some cases, writes Elton (1935, p. *xvi*), animals "choose their habitats by psychological reactions that are not very closely correlated with adaptive survival. In our own terms we should say 'they choose the habitat because they like it.'"

Fisher's (1999) and subsequent models of the runaway evolution of secondary sexual characters also assume nonadaptive preferences, but they explore only one of its many possible consequences. It was left to the self-styled evolutionary aestheticians of later generations to pursue the idea of preexisting preferences in a broader biological context.

Charles Darwin favored what we would now consider the conservative side of his era's aesthetic philosophy, exemplified by the stodgy (Browne, 1995) aesthetics of Wedgwood pottery. Citing the works of his grandfather Erasmus Darwin and of Archibald Alison, Edmund Burke, Adam Smith (as presented by Dugald Stewart), Gotthold Lessing, Sir Joshua Reynolds, and William Wordsworth, Darwin considered beauty both as perception and as emotion (Darwin, 1987). Simply stated, Darwin hypothesized that a sense of beauty and the preferences that resulted from this sense led animals to select mates independently of and sometimes in opposition to the dictates of natural selection. Falconer (1989, p. 322) refers to such tendencies as "a matter of appetite."

Some biologists and quasi-biologists holding Darwinist views do not address art in this way, but seek explanations based solely on principles of natural selection. As Ghiselin (1974, p. 130) points out, such explanations have led to considerable confusion and are based on a fundamental misunderstanding of just what Darwin meant by the term "sexual selection." In fact, "sexual selection provides instances of non-adaptive or even maladaptive characters evolving through purely reproductive competition" (Ghiselin, 1974, p. 130).

Adaptationist, often-naive views of art and literature from sociobiological and evolutionary psychological economic perspectives seem to be everywhere these days. Richardson's (2001) useful annotated bibliography offers critical commentary on many of these views as seen from the perspective of a practicing cognitive scientist. Readers interested in speculation, responsible and otherwise, about how artistic and literary creativity might in theory contribute to reproductive success and to inclusive fitness by promoting individual survival need look no further. Some of these flights of fancy attempt to use arguments based allegedly on sexual selection theory, but few of the gamesters involved actually appear to have read Darwin.

Evolutionary aesthetics in the Darwinian sense is an interesting and problematic field, if only because its history has a most intriguing human dimension. Peter Klopfer (1970), Mahdav Gadgil (1972), and Michael Ghiselin (1974) founded an evolutionary aesthetics inspired by Darwin's (1874) original, though underappreciated insights on the aesthetic component of mate selection. Darwin recognized that the inherent biases of perceptual systems would cause them to react strongly to exaggerated forms of certain sensory patterns. He further suggested that nonhumans, like humans, possess a sense of the beautiful in the form of inherent perceptual preferences. In the 1980's, theoretical models (Lande, 1981, 1982; Kirkpatrick, 1982) and experiments (Burley, 1985, 1986) broke ground for an emerging field of evolutionary aesthetics. Gould (1991) offered a charming if selective look at the first flowering of evolutionary aesthetics at the turn of the decade.

Neural Network Models, Female Preference, and Symmetry

Artificial neural networks (Anderson, 1995; Bishop, 1995; Ripley, 1996) are nonlinear devices that employ parallel distributed processing in engineering and technology applications. Perhaps their most common application is pattern recognition, but many others exist (Caudill & Butler, 1990; Nelson & Illingworth, 1991). Keith Nelson (1973) proposed conceptual neural network models of problems in animal behavior and development. Nelson discussed implications for topics ranging from female acoustic preferences to biased sensory templates for bird song, and suggested that preexisting developmental biases mediated by sensory experience could evolve analogously to the function of a neural network. Twenty years after Nelson's paper appeared, computer simulations using artificial neural networks (Arak & Enquist, 1993; Enquist & Arak, 1993) spurred a flurry of interest in evolutionary aesthetics. These simulations, and experiments by Basolo (1990a, 1990b) fuelled the evolutionary aesthetics of the 1990's. But within a few years other computer simulations (Kamo, Kubo, & Iwasa, 1998) and other experiments (Rosenthal & Evans, 1998) raised serious and still-unanswered questions about the entire enterprise. Nevertheless, interest in aesthetic preferences continued with empirical studies pursuing lines of research suggested by laboratory and field experiments with real animals and by artificial neural networks (e.g., Burley & Szymanski, 1998; Christy, 1995; Endler & McLellan, 1988; Endler, 1989; Endler & Basolo, 1998; Götmark & Ahlström, 1997; Gould et al., 1999; Ryan, 1991; Ryan, 1998). Studies by Gould and colleagues are particularly interesting because they identify dynamic movement, rather than static size or shape, as the key feature that elicits preexisting evolutionary biases in the fish they observed. These studies are also important because they involved fish that do not naturally exhibit female choice.

Some thoughtful biologists question the related concepts of nonadaptiveness and unknown origin as applied to preexisting bias. They do not question that preexisting biases exist (and even suggest the term "antecedent bias" as more appropriate), nor do they deny the evolutionary importance of such biases. However, they find the idea that such biases have unknown origin a little too mystical. They also question whether we need delve into the innards of computer models to find a possible mechanism for preexisting bias. Rather, they argue, as Darwin might have argued, that adaptive traits are hierarchical and correlated with other adaptive traits (Darwin called it "correlated variability"), and that such pleiotropic effects represent a more adequate explanation for preexisting biases than do either mysticism or computer models, while defusing the otherwise-confusing issue of nonadaptiveness. (I am indebted to Michael Ghiselin and Peter Klopfer for the thoughts that led to this discussion.)

At this writing, though the status of artificial neural network models in evolutionary theory is problematic and though Basolo's experiments remain controversial, the idea of preexisting bias rests on very solid experimental grounds, from Burley's decades of work and Ryan's studies to the paper by Gould et al. (1999). Darwin's original work and Keith Nelson's paper remain the basic source for ideas in evolutionary aesthetics. Excellent reviews (e.g., Endler & Basolo, 1998; Ryan, 1998) sought to summarize and to synthesize more recent studies in the field.

Some investigators tried to suggest that symmetry was a preferred quality of biological stimuli (e.g., Enquist & Arak, 1994, 1998; Morris, 1998; Morris & Casey, 1998; Moller, 1994; Moller & Sorci, 1998). The symmetry studies limit their scope to very simple forms of spatial symmetry for visual stimuli, as appropriate for an insect, lizard, or small bird, and disregard the generality of the mathematical definition of symmetry as invariance under transformation, as well as other preferences for asymmetric patterns in Western and especially in East Asian visual art. The entire evolutionary aesthetics of symmetry in this naive form rests on a very shaky theoretical foundation. In a telling critique of previous work on the topic, Bullock and Cliff (1997) show, using an artificial neural network model, that previous models' latent preferences for symmetry are merely artefacts of the modeling process, and that the models' alleged sensory bias requires a different explanation. These authors suggest a preference for homogeneity rather than that posited for complex symmetry. The difficulty of applying artificial neural network models to evolutionary aesthetics is even more manifest in studies of computer "preferences" for musical harmony—done by scientists (Kamo & Iwasa, 2000) whose own culture offers treasures of asymmetry and dissonance in the arts. We (both women and men) may also rightly question any mode of thought that automatically equates "female" with "receiver" and "bias." *Caveat lector*!

Internal Models and the Noise-bias Tradeoff

Computer simulations of neural network models are inherently problematic because the systems being simulated are complex, inadequately-understood, and indeterminate. But the real issue is not whether neural networks provide heuristic models for aesthetic behavior. Any artificial cognitive system, even a very simple one, will show preexisting biases like those claimed to be exhibited by some neural network models. The relevant technical domain is cybernetics, not cognitive science. For example, even a simple multiple linear regression model will exhibit a preexisting bias for linear trends in data, even when no such trends are present, when the model is implemented using a stepwise regression algorithm (Flack & Chang, 1987; Neter et al., 1996, p. 347-354). In the practice of applied statistics, it is common knowledge that stepwise regression is programmed to find linear trends and will often do so even when presented with random noise.

Kalman filter methods (Brammer & Siffling, 1989; Catlin, 1989; Harvey, 1990; Haykin, 2002; Wells, 1996; Zarchan & Musoff, 2000) seek to fit differential or difference equation models to data. These methods are also prone to preexisting biases. The phenomenon of Kalman filter divergence (Abelnour et al., 1993; Fitzgerald, 1971; Roberts et al., 1996) is a well-known drawback of these methods. The Kalman filter collects and processes observational data in real time. The filter, actually a computer implementation of a set of mathematical equations, refines and updates its current estimate of the state of an observed system, modifying the current state estimate by incorporating (in a theoretically-optimal way) newly-gathered, inevitably-noisy observations of the state of the system. Divergence results when error in model specification, poor estimates of initial conditions, or even simple computer roundoff cause the filter algorithm to rely more on its own model-based estimates and less on new observations than it should. In general systems theory, the problem is termed the "noise-bias tradeoff," and it is expected in any estimation, control, or adaptive control system with both external inputs and an internal model, as pioneer cyberneticians like Hendrik W. Bode (1945) and John W. Tukey (Blackman & Tukey, 1959) realized long ago.

The point of using a neural network model, rather than stepwise multiple linear regression or Kalman filtering, to demonstrate preexisting bias is that neural networks are superficially more similar to cognitive systems than are either multiple linear regression or Kalman filtering. Their internal biases are a result of training the network to prefer certain patterns, so that the training sequences themselves furnish the internal model that later produces the biases. The key to understanding simulation results in evolutionary aesthetics is simply to realize that any adaptive system that incorporates an internal model and that uses feedback from its surroundings to modify this model, its parameters, its control signal, or a statistic used to estimate a population parameter, is prone, and often inevitably doomed, to preexisting bias.

Do Neural Network Models Have a Future in Evolutionary Aesthetics?

Ironically, the neural network designs that stimulated the growth of evolutionary aesthetics a decade ago were technologically obsolete long before evolutionary applications of these analyses began to appear in biological journals. In the late 1980's and early 1990's, the intrinsic opacity of neural network models, i.e. the fact that the information stored in the network is difficult to extract and therefore to understand, spurred development of a new generation of so-called "neurofuzzy estimators" that combines the positive attributes of a neural network with algorithms based on fuzzy logic (Brown & Harris, 1994; Roberts et al., 1996). Paralleling these developments, a set of cybernetic techniques called "qualitative modelling" or "qualitative reasoning," based on discoveries in artificial intelligence and cognitive science, further expanded the domain of intelligent systems (e.g., Faltings, 1997; Reece, 1998; Suzuki, 1990). What is quite interesting about these newer formalisms is that they also seem to exhibit preexisting biases. They can produce spurious outcomes, inconsistent with the qualitative models underlying such formalisms (Fouche & Kuipers, 1990). You don't have to be a rocket scientist (e.g. NASA, 1993) to recognize that the study of evolutionary aesthetics might benefit greatly from updating its knowledge of cognate fields, and from looking into the possibility of a new generation of models based on neurofuzzy estimators and on qualitative modelling.

Unanticipated and Rare Events

In recognizing that animals can form expectations, Spinka et al. (2001), like Washburn & Hamburg (1965) and many others before them, hypothesized cognitive links between internal models and the environment. Spinka et al. couch their hypothesis solely in terms of a possible role for play behavior in developing the motor skills and emotional intelligence needed to deal with unanticipated physical events such as tripping, falling, or being knocked over by a big ocean wave. In a field study of lar gibbon (*Hylobates lar*) behavior, primatologist C. R. Carpenter (1940, p. 26) observed just such an event:

> The buff female of group 3 swung quickly through an open tree and out on a limb. As she was preparing to swing into an adjacent tree, the limb broke, leaving a stub which was about six inches long. As the limb broke and fell the gibbon recovered by turning almost in mid-air and catching the remaining stub of the branch. With extreme rapidity she swung around under and then on top of the limb and then, with only a slight loss of time and momentum, jumped outward and downward 30 feet to an adjacent tree top.

In recognizing that the body, like the mind, forms expectations, Spinka et al. (2001) implicitly recognize the existence of internal models, specifically in the domain of motor learning and motor skill. Geist (1978), Washburn and Hamburg (1965), and Spinka et al. focus on the behavioral and cognitive issues of predictability and anticipation, including the emotional components of perceived control or a temporary lack thereof. Their treatment invites generalization beyond the sphere of motor intelligence, indeed to all cognitive domains (a pluralistic world well-mapped by Damasio, Edelman, and Gardner) including emotion, aesthetics, creativity, and innovation. I will argue, as do Washburn and Hamburg (1965), Geist (1978), and Sutton-Smith (1997, p. 229; 2003) that play is as important for dealing with an unpredictable, uncontrollable world as it is for dealing with motor disasters ranging from a fall on icy pavement to a partner's failure to catch a dancer doing a fish dive. I believe that adaptation to unpredictability holds keys to many puzzling questions in contemporary play research, evolutionary aesthetics, cognitive science, and sociobiology.

For an event to be unanticipated, there must be an animal capable of anticipation! And anticipation is essentially prediction, based on an internal model of some kind. Though cognition is not necessarily involved, internal models must be, or prediction could not take place. A prediction has to be based on something.

Because animals are not omniscient, but still make predictions about themselves and their world, it follows that these predictions will sometimes be wrong. Indeed, the universe and everything in it may be unpredictable in a very great number of different senses. Fagen (1974), Geist (1978), Spinka et al. (2001), and Washburn & Hamburg (1965) suggested that animal play, by producing disequilibrial and vertiginous experiences, represented adaptive preparation for unanticipated events. Fisher (1999) discussed the aesthetics of rare events and the poetics of surprise from a Cartesian perspective, seeking to invoke the spirit of the Enlightenment, implicitly welcoming E. O. Wilson's (1998) call for a scientific consilience, but also, in the process, transcending these limitations to produce a work of personal sensibility, not of doctrine.

Any cybernetic device, including expert systems, which estimates or controls the state of a dynamical system using heuristics is powerless to handle the many kinds of unforeseen events that typically occur in abnormal situations (Suzuki, 1990). Model-based, qualitative reasoning using fuzzy logic can, in theory, dynamically generate new knowledge to recover from unanticipated events and to resume operations after a crash (Suzuki, 1990).

What no previous commentator seems to have recognized is that (play) behavior that evolved in an orthodox Darwinian manner to promote success in the economy of nature by preparing the individual for unanticipated events must itself inevitably lead to rare experiences, to the unanticipated and to the novel. The disequilibrial, variational structure of play sequences necessarily produces novel experience. Scratch economics and you get ludics, scratch adaptation and you get creativity and flexibility: nature pulling itself up by its bootstraps.

Emlen et al. (1998) ascribe adaptation to self-organization. Their quest seeks no farther shore. For those still inclined to look beyond the material, but willing to treat the problem in terms of multiple working hypotheses, a productive question might seem to be whether Emlen's hypothesis or the idea of intelligent design can better account for the facts of natural diversity in biological domains where play rules. These issues might well be approached purely on grounds of logic and reason, with no appeal to faith or to Biblical authority. A comparative approach considering play-based cosmological models seems as valid as any. One particular thread (or, if you prefer, string) is of particular interest. Krishna's ongoing playing is *lila*, and every moment of ongoing life is the first moment of creation (e.g., Case, 2000; Dimock, 1989; Handelman, 1992; Handelman & Shulman, 1997; Hein, 1987). Everybody into the gene pool!

Stretching Malthus

Part of the confusion that besets evolutionary aesthetics seems traceable to an overly-rigid Malthusianism. This problem confused those who accepted Darwin's economic metaphor, but failed to see how it applied to his theory of sexual selection. Quite simply, the resource in sexual selection is mates, so that sexual selection proceeds even when natural selection is inactive. A key difference, and one that by extension will illuminate two additional modes of selection beyond the natural and the eugenic, is that populations grow exponentially, so that the number of animals seeking mates and the number of mates both grow exponentially, but perhaps at different rates, leading to mate competition; whereas in Malthus, populations grow exponentially, but resources grow arithmetically. In aesthetic and ludic selection, two evolutionary modes to be further discussed here and below, resources may grow at a rate faster than exponential and are essentially infinite and inexhaustible, possibly even nondenumerable, and subject to the laws of fractal geometry, leading to mind-body rather than environment as the bottleneck in a generalized Malthusianism. Even within the narrow confines of conventional population ecology, populations ultimately limited by resources in a classical Malthusian sense can spend essentially all their time growing exponentially towards an upper limit called carrying capacity (Fowler, 1988) without sensing the resource constraint that will ultimately determine steady-state population size. Extension of Fowler's models to cyclic or chaotic population change will further enrich economic models of evolution. In fact, this scenario gives sexual, aesthetic, and ludic selection virtually free rein, with cognitive capacity, itself subject to all four modes of selection, as the ultimate limiting factor (see, e.g., Dukas, 1998) and affords the study of evolutionary ecology vastly wider scope than heretofore supposed.

Fundamental contrasts among separate factors of evolution remain foreign to many evolutionary theorists and are almost unheard-of in the work of empiricists who still find it profitable to "test" monistic hypotheses last current a decade or more ago. Darwin himself was a committed pluralist, and others have followed in his steps.

It would be far beyond the scope of this chapter to discuss alternative evolutionary realities in depth, but such realities do indeed exist and may reward a few readers. It is sobering to realize that science, like water, must take the path of least resistance in its quest for rigor and elegance, and that many phenomena, species, and even geographical regions are perforce neglected. History and politics further constrain the outcome. But we can always ask the question "What if things had been otherwise:" what if, for example, a science of evolution had been free to develop in Russia, and if, in consequence, attention had been focused on high latitudes rather than on the temperate zones and the tropics? Russian biogeographers, systematists, and evolutionists (e.g. G. E. Grum-

Grzhimailo, V. N. Potanin, I. I. Schmalhausen, L. S. Berg) and others (e.g. Geist, 1978; Gerlach & Murray, 2001; Graham et al., 1996; Guthrie, 1995, 2001; Pielou, 1988, 1991, 1994) working in periglacial environments, on reticulate evolution, and on mosaic "species" (e.g., Rice & Chapman, 1985) sometimes tend to embrace a more open concept of Darwinism than do those working in the industrialized West and at low latitudes. See also Nabokov (1989, 2000). Johnson and Coates (2001) present a history of Nabokov's studies of lepidopteran evolution and systematics from a critical but generally sympathetic perspective.

Nabokov (1962, 1989, 2000) sees a universe in which evolution works through design, though not in a literal sense, and in which an economics of plenty produced *Homo poeticus*—an interesting thought, well suited to the kinds of periglacial environments in which Geist (1978) located key steps of human evolution.

Implications of Preexisting Biases for Behavioral Ecology

That preferences of unknown origin can effectively shape evolution independently of or contrary to natural selection is not an idea to disregard, whatever the checkered history to date of its analysis by computer or animal experimentation. For example, biologists will inevitably begin to consider the implications of preferences of unknown origin in the realms of food and habitat selection, in the study of mimicry (Christy, 1995), and in the study of coevolved systems such as tree-insect (Southwood & Kennedy, 1983), plant-pollinator, plant-seed disperser, and parasite-host—in fact, "if the neural network model is correct and universal, bias should occur independently of sex, and outrageous traits should arise in both sexes, in all species, and in all network-controlled behaviors" (Heinemann, 1993, p. 308).

Implications of Preexisting Biases for Development

Keith Nelson (1973) proposed that a developmental template modifiable by experience could exhibit bias. His discussion of bird song learning generalizes immediately to all experience that has long-term developmental effects on the phenotype. Preexisting bias for experience will feed back into all developmental templates that convert that experience into phenotype. If this preexisting bias falls within the domain of aesthetic factors rather than the economic or the eugenic, the arena of development broadens substantially. We see preexisting biases that produce experiences that modify templates that steer development in novel directions. All of this places the entire program of genetic and epigenetic determinism so familiar to present-day biology watchers in serious question. These scientific ideas and their rich history should prove particularly accessible to scholars in the humanities and social sciences, and especially to those seeking maps of the borderlands joining contemporary biological science with their own fields. Kortmulder (1998) and Goodwin (1994) identify some of the key directions taken by biologists interested in the problems of emerging order and design in nature. For a yet-broader overview of these ideas in a global context

linking animal play and T. S. Eliot with elements of classical Indian philosophy, please consult *Springing the Trap of Aesthetic Evolution: Animal Play and
the T. S. Eliot Effect* below and the Conclusions section of this chapter.

Cognitive Preferences and Transformations

To begin filling one obvious blank area on the map of evolutionary aesthetics, we must consider cognitive as well as sensory preferences: preferences for
imagined colors or smells, for imagined patterns of sensation or sensations, and
preferences of unknown origin for ideas themselves, particularly in the case of
humans, their close primate relatives, and other large-brained species not closely
related to humans. Human nature in its Proustian sense is a complex amalgam
of emotion and memory in Bergsonian time. As such, it can and readily does
transform and transcend hypothesized "universal" standards of human attractiveness (e.g., Enquist & Ghirlanda, 1998; Etcoff, 1999). In Proust's work, perhaps
the most illuminating discussion of these transformational processes appears in
Albertine disparue (Proust, 1989, p. 21-24). In this passage, Marcel, the narra-
tor of the work, contrasts the immense bile, chagrin and folly that a great and
tragic love can cause with the unprepossessing appearance of the loved one's
photograph in the eyes of an unbiased observer. That the biases in question are
indeed not universal, but contingent stems from the fact that "the construction
of sensations interposed between the woman's face and the lover's eyes, the vast
mournful egg that envelops and misrepresents her as a layer of snow would a
fountain, are already extended so far that the point at which the lover's gaze
stops, the point at which he finds his joy and his suffering, are as far from the
point at which other people see her as the real sun is from the place in the sky
where its condensed light makes it visible" (Proust, 1989, p. 22; my translation).

I believe that, particularly where humans are concerned, the aesthetics of
sensory bias and preexisting preferences undergo fundamental, revolutionary
change under the influence of mind: a distinction analogous to, but more far-
reaching and comprehensive than that made in classical psychology between
perception and sensation. In classical poetics and aesthetics, this transformative
interpenetration represents an example of the so-called "T. S. Eliot effect"
(Kubler, 1962, p. 35 and n. 4), which will play a large role in the subsequent
discussion (*Springing the Trap of Aesthetic Evolution: Animal Play and the T.
S. Eliot Effect* below). Evolutionary aesthetics may well unveil a novel biological world in its studies of fish and frogs and finches, but the newness of this
world is, I predict, trivial compared to that which will emerge from a proper
appreciation of the transformative role of mind in aesthetic evolution, and in
particular in human evolution. "Evolutionary psychology" is, in fact, a science
yet unborn.

Dislike, distaste and disapproval (Darwin, 1987, p. 587) are the other side
of the coin of evolutionary aesthetics. Darwin himself realized this, of course,
but the field of evolutionary aesthetics seems skewed towards likes rather than
dislikes. The dark places of the human mind fascinated Freud and Jung, and
surely must have far more to offer biologists than evolutionary psychology,

with its single-mindedly Panglossian adaptationism, has so far envisaged. We need only read Kafka to get a true sense of these vast dark spaces.

Did Someone Mention Beauty?

I question the immediate and uncritical equating of sensory preference and sensory bias with beauty. Pop sociobiology and evolutionary psychology have long been guilty of making hasty conflations of this sort. Such flawed logic is of service to no one. The hard-core pervert straining after Internet images is drawn, allured and grabbed, but are grotesque bodies and grossly-magnified anatomical images beautiful, and is this stance contemplative in Sparshott's sense? Perhaps in a more sophisticated cultural domain this might be the case, but surely not in the banal universe of adult websites. Does aesthetics even apply here? And what of food that might seem attractive to a hungry person, but that would not meet even the most elementary standards in a cuisine that values presentation and visual appeal (French, Japanese, Chinese, Thai)? There is need to apply critical thought to the distinction between beauty and sensory preference as it occurs (or should occur) in evolutionary aesthetics, and to address this issue in a searching fashion by using the hardened and unsentimental tools of philosophy. Any treatment of evolutionary aesthetics that purports to represent bona fide aesthetic philosophy needs to measure up to Sparshott's (1998) definition of the contemplative stance in terms of disinterested attention. Legitimate biologies of art are clearly possible in theory, but not yet, it seems, in fact.

Springing the Trap of Aesthetic Evolution: Animal Play and the T.S. Eliot Effect

Animal play behavior makes more sense as an antidote to aesthetic drive than as a component of it. Suppose that the capability to play appeared at a low frequency in a population of animals with preexisting sensory biases. If these biases were maladaptive but persisted in evolutionary time because of a stable balance between natural and aesthetic selection, could not play evolve by counteracting or even by reversing the effects of these biases? Animal play in general, and human play in particular, offers two classes of mechanism that can spring the traps set by aesthetic evolution.

Class 1: Animal play can tune, switch among, replace, or reshape preexisting biological tendencies. Thus, for example, it can reshape the fabric of emotion by injecting intelligence, by adjusting the shapes and time constants of emotional norms of reaction, or by decreasing the organism's attachment to its internal models or confidence in their validity, in a safe context free from economic stressors. In Kalman filter design, by analogy, a common cure for divergence is to add driving noise to the filter's estimated covariance matrix. This mindless expedient tells the filter to place less weight on its previous estimates of the state of the system and more weight on new observations when forming an updated estimate. Because the filter's confidence in its own estimates is thus artificially decreased on every update of the state estimate, it is more open to new information. In animal behavior, where the action involves entire internal models rather than just their parameters or the state estimates based on them, we may speak in terms, for example, of switching from one behavioral attractor's basin to another, or from a universe defined by one collection of attractors to an alternative (biologically dependent) reality defined by an alternative collection of attractors.

Class 2: Animal play can create novel, emergent patterns of perception, emotion, cognition, and behavior in three ways.

The first way involves consequences of play unintended by natural selection, under the assumption that play is adaptive preparation for unpredictability (Fagen, 1974; Geist, 1978; Spinka et al., 2001, Washburn & Hamburg, 1965). Play that evolved to function as preparation for rare, unpredictable events must necessarily have a behavioral structure that includes novel, unusual, and idiosyncratic postures, movements and movement qualities, gaits, phrasings, rhythms, and relationships with features of the enviroment and conspecifics (Fagen, 1974; Spinka et al., 2001). The result of this behavior is novel experience, and the result of the novel experience can itself be a novel and emergent phenotype, far more versatile than an animal that is merely a little bit better prepared for a few more kinds of emergencies. In fact, the simple fact of emergency preparedness makes an animal capable of moving into new environments, new modes of thought and feeling, and new adaptive zones. Perhaps, since play is intrinsically rewarding, play designed as preparation for the unpredictable can produce a taste for the unpredictable and novel. In turn, the formation of what is essentially an aesthetic preference for novelty and unpredictability as a result of play experience can be tuned by play so that it is no longer maladaptive. This process has no limits.

The second way invokes a metaphor and the spirit of qualitative modelling. (To cite Lakoff and Johnson (1980) here may be uncritical, in that it is philosophically incorrect to extend the use of the term "metaphor" beyond the domain of figures of speech). The domain of discourse is the study of dynamic systems, chaos, fractals, and self-organization. Play can create novelty via a process akin to self-organization in complex systems. Invoking dynamic models that contain mathematical objects known as attractors, Emlen et al. (1998) emphasize the unappreciated power of self-organization to produce novel adaptive patterns in nature. Several authors, including especially Kortmulder (1998) and Goodwin (1994), view play in this light. If you imagine an individual's behavioral development or stream of ongoing behavior across time as a turbulent stream with many local eddies, and each eddy as a local attractor, that is, a specific behavior pattern that tends to repeat itself, then the individual can be like Winnie the Pooh's stick, or Eeyore, potentially trapped and going around and around forever, unless something can move it out of the potential well represented by a local basin of attraction.

Many possible factors can cause behavioral change. Play would be one. It would loosen the grip of one attractor. Then, the animal, or, more correctly, the state vector describing its current behavior, would shift to the domain of another attractor. In the traditional Markov chain model of behavior, the state vector has a finite number of elements, one for each behavior type in the ethogram. But behavioral elements can be far more numerous than are the integers. Perhaps the true dimensionality of behavior, particularly if we also include attributes like quality, rhythm, phrasing, and transitions in our models (Bartenieff, 1980; Dell, 1977) is fractal, uncountable, or even transfinitely infinite. See, e.g., Burrill & Knudsen (1969) for discussions of concepts of numerosity and of transfinite numbers.

Clearly, there is nothing to keep us from moving up in the hierarchy of levels of organization and saying that play could do more than simply changing the animal's state. It could go farther and deeper, and change the entire dynamic regime. This would be equivalent to a change, while travelling downstream, from pool to riffle to glide, rather than just a change from one eddy to another in a single riffle. At this higher, more abstract level, we are not simply speaking of a change in the value of state or of parameter values (i.e., a change from one behavior type to another, or a change in the values of one or more of the elements of the transition probability matrix of a behavioral Markov chain model). And, indeed, when such a global change in dynamic regime takes place, other transformations necessarily follow. This revolutionary sort of change is the subject of the T. S. Eliot effect, which I choose to identify as the third of three ways in which play can create novel, emergent patterns of sensation, perception, cognition, emotion, and behavior.

To make their science agreeable, behavioral ecologists and evolutionary psychologists have, on more than one occasion, resorted to handy, portable catch names. Students, the media, and the general public relate readily to tags like Dawkins' (1976) "green beard effect" or Krebs' (1977) Beau Geste. Likewise, in evolutionary aesthetics, the T. S. Eliot effect embodies an entire scientific direction. I will seek to show that it also deserves top billing on the playbill of contemporary behavioral biology.

Every major work of art forces upon us a reassessment of all previous works. This view defines the T.S. Eliot effect (Kubler, 1962, p. 35, n. 4; Steadman, 1979, p. 238; see also Gombrich, 1961, p. 27). Everywhere in biology and in the humanities, and particularly in the study of art and play, we see novelty emerging by the same analogous process. Every epoch acquires fresh eyes (Praz, 1967, p. 34). There is nothing that may not be new, nothing that may not recover its pristine freshness in our consciousness (Valéry, 1964, p. 115). As human society arises out of animal societies, it transforms the adaptations it possesses, therefore making new needs (Lewontin, 1977). The T. S. Eliot effect offers new views of human nature and of life's history on earth.

By the T. S. Eliot effect, the phenotypic novelty produced by play can re-invent the animal's world as well. In an ecological model, we would see changes in the numerical values of the parameters and state variables that quantified the animal's estimates of the environmental variables that had some importance for its survival and reproduction. In a simple, non-cognitive model of behavior, we would also see changes in these values for the animal's own phenotype and phenotypic estimates. In general, we would see changes in the state parameters, parameters, and equations or qualitative rules that governed the animal's internal models of itself, of its world, and of its interactions. As Trivers (1985) long ago argued, "biological dependent variables" such as these are the rule in nature–a rule that many ecological modellers, particularly those in applied areas, seem loath to recognize.

The result of the T. S. Eliot effect may be implemented by adding driving noise to produce a decrease in the animal's confidence in its internal models at an appropriate level in the modelling hierarchy. But the effect itself is not the *deus ex machina* of driving noise mindlessly added, in behavioral time by mechanisms shaped by selection, to some hypothetical or metaphorical covariance matrix, or similarly subtracted from some hypothetical or metaphorical Fisher information matrix (driving noise genes! totally awesome! [sarcasm intended, sorry]) but something more akin to the change in perception and cognition inspired by a work of art, however simple (anything from the celebrated cup/faces {Rubin's cup} or rabbit/face drawings to the experience of seeing what cannot be unseen in a Rembrandt painting or Dickinson poem, and thereafter in the world).

Of course, we humans find many aspects of animal and human play aesthetically appealing. But much animal behavior has aesthetic appeal, and we should no more focus on play because of its apparent goallessness than we should focus on the natural grace of any young organism as the answer to problems of evolutionary thought. Instead, let's look at the artlessness of play. We are not looking at a polished, rehearsed performance. Play is unstudied. The next time you have the chance, if you do, watch ballet students horsing around. Compare what you see with what they do in class.

We are much closer now than we were in the 1970's to understanding the links between the structure, or lack of structure, and function, or lack of function, of animal play. Spinka et al. (2001) refined, updated, and helpfully operationalized some hopeful suggestions (Fagen, 1974) on structure and function in animal play. The biology of animal play offers clear guidance here as to exactly what it is that we are seeing. Experiments on laboratory rats in restricted environments (Einon & Potegal, 1991; Potegal & Einon, 1989) show that play experience is not practice or rehearsal of the motor skills needed in future fighting. Rat play has specific effects on emotional intelligence, and particularly on the ability to judge just what is threatening in a given situation and when to behave defensively. Einon and Potegal's experiments strongly suggest a role for laboratory rat social play in imparting emotional intelligence to agonistic behavioral domains near the classical locus of defensive aggression. Spinka et al. (2001), appropriately citing the Einon-Potegal studies, try to suggest just how behavior with the structure of social play should reveal specific behavioral structures (scaffolding?) designed by evolution to develop emotional intelligence in particular sectors of the agonistic domain.

To interpret Einon and Potegal's (1989, 1991) important findings in the simplest possible terms, play (in rats, in these experiments) makes emotional intelligence develop, and by doing so makes the player behaviorally and cognitively adaptable, flexible, resilient, and versatile. Adaptability and versatility are the issues, not motor skill. To be wise rather than impulsive or mindlessly reactive, it takes the ability to unhook one's self from the here and now, whether the experience is one of environment or mind. Indeed, emotional intelligence needs to be an unhooking from an entire dialectic - Levi-Strauss (1992, p. 414) called it "unhitching." Since whether or not something is a threat depends on everything from how close the nearest hiding place is to how well fed you are, this is not an invariant stimulus-response learning paradigm but something more nearly akin to an ongoing assessment of one's overall well-being and quality of life. Everybody might be beautiful at the ballet, but it's not true that everybody is beautiful on the playground. Unhitching through play is the royal road to freedom from sensory-cognitive biases of unknown origin. Though the role of play in emotional intelligence was first demonstrated in the context of defensive aggression by Potegal and Einon's (1989) work, the scope of this approach to play is far broader, and applies without loss of generality to nonsocial play forms in which animals derive experience from body movements or from interactions with objects or the landscape. In these forms, play modulates blind fear and blind trust, and produces more flexible forms of predatory and predator avoidance behavior, as well as emotionally intelligent behavior specifically relating to intraspecific aggression.

Just as sensory (and cognitive!) preferences of unknown origin added a new dimension to evolution, perhaps as early as the evolution of the first vertebrates, the evolution of play added yet another dimension along which the operation of selective pressures in the natural and aesthetic dimensions could be reversed, counteracted, or overruled. This transformation, this transcendence, is precisely analogous to that which in Proust's treatment of love contrasts the lover's view of his beloved with an unbiased observer's view of her unimpressive photographic image.

Play might have evolved in primitive mammals during the early Mesozoic, about 240 million years before the present. The age of the great dinosaurs was still thirty milliion years in the future, and the first birds would not appear until the end of the Jurassic, almost a hundred million years away. Play must have evolved independently in the mammals and in the birds, the two major taxa in which play behavior is widespread in extant species. But possible cognate developments in other vertebrate groups (dinosaurs apart from birds? crocodiles and alligators, which form the third major branch of the archosaurs, apart from birds and dinosaurs?) continue to inspire speculation by biologists.

By the time humans appeared, play had a long evolutionary history, and aesthetics went back much farther, perhaps to the first vertebrates, or possibly even earlier. Both play and aesthetics had been operating during much of the long evolutionary history of mammals and birds. We can conclude that any biologically-based explanation of human behavior that only considers Darwinian natural selection, or one that considers natural and sexual selection without considering aesthetic components of sexual selection and nonadaptive biases, or one that considers natural and aesthetic selection without considering play, is automatically questionable.

Beleaguered play scholars and aestheticians suffering from information overload: if you have managed to read this far, you may deserve to be numbered among the few true heroes on this planet. Evolutionary psychologists and sociobiologists have sought to undermine your life's work, postmodernism has continued to do what it does best from an entirely different direction, and here you are reading *Play and Culture Studies* and wondering if you'll even manage to finish this volume before the next one appears. What use is a wordy critique of preexisting cognitive biases towards genetic determinism and towards materialism when we could be reading Huizinga, Victor Turner, Gregory Bateson, Don Handelman, or Brian Sutton-Smith? But remember that the brew of evolutionary aesthetics is still inchoate, controversial, full of pitfalls, and well-spiked with the heady wine of scholarly self-interest —a not-uncommon state for a new discipline. Surely, the most exciting insights lie ahead.

Play is artless and unstudied, the source of the creative spirit rather than the shaper of artistic products. But play is also one of evolution's five principal gateways. These five gateways lead to five evolutionary roads through the landscape of variation in nature: the economic, the eugenic, the aesthetic, the ludic, and the agapic. These dimensions of evolution are independent and coequal. This pluralistic approach allows what may be the most complete and general conception of evolution to date, one of the first truly substantive generalizations and extensions of Darwin's work.

Darwin situated a historical dimension of human nature in phylogeny. In doing so, he altered the concept of human history in a radical, revolutionary way. Because both bodies and minds have a linked phylogenetic history, it is possible to view perception, emotion and cognition as biological dependent variables. Biologists interested in human nature do not dispute this point, but they disagree fundamentally regarding its interpretation (see, e.g., the ongoing dispute about possible biological and cultural bases of female beauty: Manning, Trivers, Singh, & Thornhill, 1999; Toveé & Cornelissen, 1999; Yu & Shepard, 1999). In the following paragraphs, I discuss the biology of mind-body linkages and the aesthetics of emotion.

In the world of perception, touch (Josipovici, 1996; Tuan, 1974, p. 7), pressure, and sensed movement deserve pride of place (Klopfer, 1970; Sacks, 1998; Sheets-Johnstone, 1990, 1999). Bodies move, and mind and thought are a specialized form of motion. In choreographer Yvonne Rainer's words, the mind is a muscle (Lakoff & Johnson, 1999; Sheets-Johnstone, 1990, 1999) whose activity resides fundamentally in the linked functions of metaphor or "metaphor"(if Lakoff & Johnson, 1999 can be believed) and quality (Goodwin, 1994; Sheets-Johnstone, 1990, 1999). In health, body and mind move as one; this characterization may well serve as a definition of health. Suzanne Berger's *Horizontal Woman,* (Berger, 1996), a powerful first-person account of a disability that prevented her from walking or sitting for many years, and other histories of physical disability and impairment (e.g., Sacks, 1998) clearly demonstrate the suffering that can occur when mind and body disconnect.

The embodied mind holds within itself the entire treasure house of organic evolution, and every movement, whether new or phylogenetically ancient, is a jewel from some biological treasure chest. Our bodies encompass the encyclopedia of life's evolution, and with training can, like dancers (Sparshott, 1995), open it to any page. Whether certain pages or even volumes may be missing remains to be determined. In non-Western cultures, training in mindful movement and in internal martial arts extends the scope of movement as living history to the inanimate world, equivalencing human ribs to the ribs of the mountains, and breath and circulation to the flow of air and water. The body's access to the entire universe stops only at the spatiotemporal limits of the universe itself. How heavy is tradition? Its weight is equal to the total mass of the universe. Can there be anything new under the sun? And if not, as Steven Jay Gould once suggested, can we rely on the manifold powers of combinatorics to work unceasing wonders?

Aesthetic emotions are primary. In the psychology of mind (Sutton-Smith, 2003), as in theoretical and applied animal behavior (Hearne, 1986; Masson & McCarthy, 1995) ideas about emotion are again current. Emotion is body, mind, and much more, as Hinde (1985) pointed out long ago in his prescient and penetrating review of the subject. In Darwin's studies of mind and metaphysics, thoughts on beauty and emotion were pivotal, particularly in the theory of sexual selection by female choice (Ghiselin, 1974). But current studies in the evolutionary psychology of aesthetics take a fundamentally different approach from Darwin's. They leave emotion out, and base their conclusions on experimentally-demonstrated preferences shown by confined animals or by computer models. It remains to be seen whether this reductionist approach will ultimately be successful, as has been the case, for example, in the study of behavioral ecology, or whether it will be necessary to start again from Darwin's expressed view that emotions are as essential to animals' sense of beauty as are sensory preferences of unknown origin.

Mind is body, body mind. Aesthetic emotions are primary. From these bases, evolution and its cultural cognates shape the internal alchemy of desire.

Love Deconstructs Beauty

Ghiselin (1974), based on close readings of Darwin, discusses evolution in terms of an economy of nature. A view of evolution consonant with perspectives based on aesthetics would take the idea of a poetic economy (Valéry, 1964) of nature as its starting point. This view helps explain why "Beauty is in the eye of the beholder" and "Beauty is only skin-deep" are two important biological laws. (Is a thing of beauty a joy forever? Is love blind?)

Steadman (1979, p. 238), discussing the T. S. Eliot effect, argues strongly against "the belief that regularities of a reproducible and universal, presumably biologically-based, kind may be determined in the behavioural or aesthetic responses which people make to certain architectural forms, spatial arrangements, uses of colour and so on." In their brief critique of previous work in evolutionary aesthetics, Yu and Shepard (1998, p. 216) similarly argue that "cultural invariance combined with an adaptive story is fraudulent." An evolutionary aesthetics based on the preferences and responses of fish, birds, and computers seems inadequate to the task of explaining what we can observe or infer of mind and behavior in animals who lead rich emotional lives.

In Proust, memory transforms emotion and emotion memory, and human behavior emerges from the interplay of mind and desire. The sense of beauty conveys fleeting impressions that only persist when transformed by some other association, emotion or thought: for example, the significance of a beautiful place in history or in one's past life (Tuan, 1974, pgs. 93-95). Tederheid (Kortmulder, 1998, pgs. 27-42), a kind of caring tenderness made manifest by its four qualities of softness, tangentiality (slantness or obliqueness), inwardness, and mutuality, is clearly a primary force in the transformation of aesthetic emotions based on preexisting perceptual biases. A behavioral biology that treats aesthetic factors as an independent evolutionary mover, equally-orthodox to economic and eugenic factors, is surely an improvement over one that does not. But any species that can experience tederheid is no longer equivalent to a computer-based neural network model with preexisting perceptual biases. As Kortmulder (1998, pgs. 30-31) argues, tederheid is a key element of behavior even in a fish, the Paradise Fish *Macropodus opercularis*. Quite possibly it will be found to be widespread even in the "lower" vertebrates.

A girl growing up in a society whose standards of beauty are defined by supermodels and Disney fairytale princesses may lack any sense of self-worth by the time she is a teen-ager. If evoutionary aesthetics purports to show that a preference for supermodels and Disney fairytale princesses is in our genes, it has a lot to explain. Supermodels and Disney princesses seem designed, whether intelligently or otherwise, to give most girls low self-esteem. Perhaps the same could be said of "in our genes" evolutionary aesthetics. How often do people fail to believe in themselves and not realize how important they have become to another? Are there fairytale endings even for those excluded from Disney's fairytale kingdom and from the high castle of beautiful genes? Perhaps. Given any preexisting perceptual bias, we can always hope that tederheid is waiting in the wings to transform it. Love can conquer all.

Play establishes, renews and protects love shaped by tederheid from beauty. Whether we consider courtship play in nonhuman mammals, two counselors at a summer camp building a tower of milk cartons between them in the best Winnicottian tradition of a transitional object, now not between mother and young but between two young adults growing ever-closer, we see that play lends the emotion of love a kind of intrinsic intelligence, doing whatever it can to keep emotion from being dominating, controlling, exploitative, or selfish, and helping to ensure that Paul's words on love to the Corinthians will not fall on deaf ears. The mechanism is very simple, a straightforward application of play as the prime mover of emotional intelligence and resilience in Einon and Potegal's (1989, 1991) experiments, but it makes the world go round.

There are some intriguing hints that play alone may not be enough to ensure harmony and balance. Play is in a sense amoral, though it too has rules of good behavior (self-handicapping, for example). More than a half-century ago, Elizabeth Sewell (1952) devised an original approach to the nonsensical use of language and developed profound insights into the moral contexts within (or outside) which both play and nonsense operate. Her treatment of play, I belatedly learned, remains to be discovered by other play researchers, and is, I consider, of the highest importance in understanding both human and nonhuman play.

Predators' play with prey, if it can even be considered play (I believe it can, but others disagree) seemed to Darwin an example of the seeming cruelty of nature. I don't think that play has cruel intentions, but it can sometimes have cruel consequences. For this reason, and because the theory of reciprocal altruism still falls short of explaining much in nature that might seem truly altruistic, the theoretician is left with a conundrum to solve. Do we try to extend reciprocal altruism, do we try to stretch play thin enough to cover phenomena that perhaps it was never meant to address, or do we expand the scope of artificial selection in nature to allow love (agapic factors, to continue the Greek nomenclature) equal standing with play and beauty? There may well be five gates, the fifth being love. It's only a hunch, but see, for example, Ghiselin's (1974) discussion of play and artificial selection, in which aesthetic, ludic, and agapic factors all appear. A parallel insight, at least in spirit, may be present in G. E. Moore's suggestion (in the *Principia Ethica*, 1903) that only personal relations and beautiful things are of fundamental value.

CONCLUSIONS

This long chapter reaches conclusions that a few words can summarize. Play, emerging from aesthetic evolution as aesthetics emerged from the economy of nature, utterly transforms what it finds, while love waits in the wings to transform the transformation. Charles Darwin knew (see also Darwin, 1874, p. 939, and Ghiselin, 1974, p. 258):

> Seeing a puppy playing cannot doubt that they have free will. (Darwin, 1987, p. 536)
> [C]at pounces & runs after feather, it knows it is not mouse, but does it not use imagination & picture to itself it is. (Darwin, 1987, p. 585)

A pluralistic view of evolutionary factors opens unanticipated vistas. These perspectives can produce considerable clarity when applied to the problems of evolutionary history and human nature that sociobiology, evolutionary ecology and cognitive science have sought to address. As such, evolutionary aesthetics has manifold implications for society at large.

We can take (faint) hope in the fact that the flaws in the first generation of experiments and models in evolutionary aesthetics did not take long to surface. Because evolutionary aestheticians are intrinsically interested in discovering and understanding intrinsic biases, whereas engineers, programmers, and managers tend to view them as annoyances to cover up or to fix as unobtrusively as possible without stopping to understand what really went wrong, it could be that new discoveries in evolutionary aesthetics will be the first to shed light on the intrinsic biases of large, complex, intelligent systems in general. As such, these discoveries would represent a major contribution of Darwinism to society.

The fivefold gates of evolution make it possible to think about art from a significantly-new biological perspective. The way of economics shows us that in theory, animals can be selected to prefer, and perhaps to view as beautiful, that region of resource space which tends to increase their individual Darwinian fitness. The way of eugenics shows us that in theory, animals can be selected to prefer, and perhaps to view as beautiful those mates whose genes will enhance the survival of the subject's own genes for the preference. The way of aesthetics shows us that in theory, animals can be selected to prefer, and perhaps to view as beautiful those stimuli, perceptions, sensations, ideas, experiences, and regions of resource space for which the subject has a preexisting perceptual-cognitive bias. The way of ludics shows us that in theory, animals can be selected to perform behavior called play which relaxes the constraints imposed by the ways of economics, eugenics, and aesthetics. The way of agapics is as profound as beauty is skin-deep, adds compassion to play's brittle brightness, and shows us the power of artificial selection at work in nature.

Natural order can be ornamental, can reveal meaningful and pleasing artistry, and can even inspire (faint) hope. Aesthetics and play constitute human nature and the nature of those animals most like us phylogenetically or ecologically. Under the aegis of love, intelligent design emerges in a process of ongoing creation. The challenge of being human is to find ways to participate more fully in this process.

ACKNOWLEDGEMENTS

Brian Sutton-Smith and Michael Ghiselin instigated this chapter. I thank them for their encouragement and patience. Of course, they can't be blamed for the results! Peter Klopfer and Koenraad Kortmulder helped by supplying key references at a crucial time and by affirming the continuing value of pluralistic approaches to animal behavior and to human nature.

Especial thanks to Roger Lathbury for introducing me to Elizabeth Sewell's work. And thanks to Jack Piccolo and Nick Hughes for suggesting that I read Elton. Thanks to Ghiselin, Klopfer, Kortmulder, and Sutton-Smith for comments on earlier drafts. I only wish there had been more time during the eventful year of 2001 to read all the literature that they suggested and to pursue every original idea that each of these four readers generously offered. I thank Stephen Mo Hanan for allowing me to use the title of his wonderful comedy *Everybody Into the Gene Pool* as a section heading, and Brynar Mehl for helping me better understand classical Indian philosophy and the ongoing play of creation through many fruitful discussions and artistic work in the Cecchetti-Craske tradition of theatrical dance. Finally, I am grateful to Philip Kitcher for posting some very useful words by G. E. Moore on his excellent website (visit http://www.columbia.edu/~psk16/ for this and other cool stuff).

I was trained as a Harvard sociobiologist, but reneged. I acknowledge the clarity, order, richness, and rare but welcome humor of many studies in the biology of social behavior, and particularly of works by W. D. Hamilton, Robert Trivers, and E. O. Wilson. Though I now disagree with many aspects of the sociobiological program and with many if not most of its executors in evolutionary psychology, I cannot deny that from a writer's perspective, Wilson's books will long endure as models of masterful scientific prose, just as has been the case with the works of another master writer, D'Arcy Wentworth Thompson, arguably the greatest prose stylist in all of biology.

Seven years ago Michael Ghiselin introduced me to pluralistic evolutionary factors and to problems of evolutionary aesthetics, and he suggested that I write a review on the topic. In the early summer of 2001, Brian Sutton-Smith casually asked me if I had anything to say currently about play and art. He read and commented on the resulting draft, and he sent copies of recent works in evolutionary psychology, including pre-publication copies of his papers on the psychology of emotion and play.

Students in the Juneau Choreography Project offered thoughtful comments, lynx-eyed scrutiny, and a guide to the world of *temps perdu*. Thanks also to Whitney Harding and to Patti Mattison and her students, especially Emily D., Jessica D., Lael R., Emily S., Marissa C., Rebecca S., Shailyn H., and Chris S. for helping me survive the final draft.

REFERENCES

Abelnour, G., Chand, S., Chiu, S. & Kido, T. (1993). On-line detection and correction of Kalman filter divergence by fuzzy logic. *Proceedings of the American Control Conference, San Francisco, CA, 1835-1839.* absolutely. (2000). Vanessa's webpage. URL: http://www.gurlpages/absolutely/

Anderson, J. A. (1995). *An introduction to neural networks.* Cambridge, MA: MIT Press.

Anderson, R. R. (2001). *Lynx Comics; Zoz Comics.* URL: http://www.holisticforgeworks.com.

Arak, A. & Enquist, M. (1993). Hidden preferences and the evolution of signals. *Philosophical Transactions of the Royal Society of London Series B: Biological Sciences, 340,* 207-213.

Bartenieff, I. with D. Lewis. (1980). *Body movement: Coping with the environment.* New York: Gordon and Breach Science Publishers.

Basolo, A. (1990a). Female preference for male sword length in the green swordtail, *Xiphophorus helleri* (Pisces: Poeciliidae). *Animal Behaviour, 40,* 332-339.

Basolo, A. (1990b). Female preference predates the evolution of the sword in swordtail fish. *Science, 250,* 808-810.

Bauer, H. R. (1976). Discoveries with Jane Goodall at Gombe Stream. Seminar presented at University of Illinois, Urbana-Champaign.

Berger, S. (1996). *Horizontal woman: The story of a body in exile.* Boston: Houghton Mifflin.

Bishop, C. M. (1995). *Neural networks for pattern recognition.* Oxford: Clarendon Press.

Blackman, R. B. & Tukey, J. W. (1959). *The measurement of power spectra from the point of view of communications engineering.* New York: Dover.

Bode, H. W. (1945). *Network analysis and feedback amplifier design.* New York: Van Nostrand.

Brammer, K. & Siffling, G. (1989). *Kalman-Bucy filters.* Norwood, MA: Arteh House.

Brand, P. Z. (2000). *Beauty matters.* Bloomington, IN: Indiana University Press.

Brown, M. & Harris, C. J. (1994). Multi-sensor data fusion for obstacle tracking using neurofuzzy estimation algorithms. *SPIE Optical Engineering in Aerospace Sensing, Orlando, FL, 2223.*

Browne, J. (1995). *Charles Darwin. v. 1. Voyaging.* New York: Knopf.

Bullock, S. & Cliff, D. (1997). The role of 'hidden preferences' in the artificial co-evolution of symmetrical signals. *Proceedings of the Royal Society of London B: Biological Sciences, 264,* 505-511.

Burley, N. (1985). The organization of behavior and the evolution of sexually selected traits. In P. A. Gowaty & D. A. Mock (Eds.), *Avian monogamy* (pp. 22-44). Ornithology Monograph 38, American Ornithologists Union.

Burley, N. (1986). Sexual selection for aesthetic traits in species with biparental care. *American Naturalist, 127,* 415-445.

Burley, N. & Szymanski, R. (1998). A taste for the beautiful: Latent aesthetic mate preferences for white crests in two species of Australian grassfinches. *American Naturalist, 152,* 792-802.

Burrill, C. W. & Knudsen, J. R. (1969). *Real variables.* New York: Holt, Rinehart & Winston.

Carpenter, C. R. (1940). A field study in Siam of the behavior and social relations of the gibbon (*Hylobates lar*). *Comparative Psychology Monographs, 16,* 1-212.

Case, M. H. (2000). *Seeing Krishna.* New York: Oxford University Press.

Catlin, D. E. (1989). *Estimation, control, and the discrete Kalman filter.* New York: Springer-Verlag.

Caudill, M. & Butler, C. (1990). *Naturally intelligent systems.* Cambridge, MA: MIT Press.

Christy, J. H. (1995). Mimicry, mate choice and the sensory trap. *American Naturalist, 146,* 171-181.

Clark, M. (2000). *Revenge of the aesthetic.* Berkeley: University of California Press.

Darwin, C. (1874). *The descent of man, and selection in relation to sex.* 2nd ed. London: J. Murray.

Darwin, C. (1987). *Charles Darwin's notebooks, 1836-1844:* Geology, Transmutation of Species, Metaphysical Enquiries. Barrett, P.H., Gautrey, P. J, Herbert, S., Kohn, D., & Smith, S. (Eds.). London: British Museum (Natural History), and Ithaca, NY: Cornell University Press.

Dawkins, R. (1976). The selfish gene. Oxford: Oxford University Press.

Dell, C. (1977). *A primer for movement description.* Revised edition. New York: Dance Notation Bureau Press.

Dewey, J. (1934). *Art as experience.* New York: Minton, Balch & Co.

Diamond, J. (1986). Biology of birds of paradise and bowerbirds. *Annual Review of Ecology and Systematics, 17,* 17-37.

Dimock, E. C. (1989). Lila. *History of Religions, 29,* 159-173.

Dissanayake, E. (1992). *Homo aestheticus: Where art comes from and why.* Seattle, WA: University of Washington Press.

Dukas, R., (Ed.). (1998). *Cognitive ecology.* Chicago, IL: University of Chicago Press.

Einon, D. & Potegal, M. (1991). Enhanced defense in adult rats deprived of playfighting experience as juveniles. *Aggressive Behavior, 17,* 27-40.

Elliott, E., Caton, L. F. & Rhyne, J. (Eds.). (2001). *Aesthetics in a multicultural age.* New York: Oxford University Press.

Elton, C. (1930). *Animal ecology and evolution.* Oxford: Oxford University Press.

Elton, C. (1935). *Animal ecology.* 2nd impression. London: Sidgwick & Jackson.

Emlen, J. M., Freeman, D. C., Mills, A. & Graham, J. H. (1998). How organisms do the right thing: The attractor hypothesis. *Chaos, 8,* 717-726.

Endler, J. A. (1989). Conceptual and other problems in speciation. In D. Otte & J. A. Endler (Eds.), *Speciation and its consequences,* (pp. 625-648). Sunderland, MA: Sinauer.

Endler, J. A. & Basolo, A. L. (1998). Sensory ecology, receiver biases and sexual selection. *Trends in Ecology and Evolution, 13,* 415-420.

Endler, J. A. & McLellan, T. (1998). The process of evolution: Towards a newer synthesis. *Annual Review of Ecology and Systematics, 19,* 395-421.

Enquist, M. & Arak, A. (1993). Selection of exaggerated male traits by female aesthetic senses. *Nature, 361,* 446-448.

Enquist, M. & Arak, A. (1994). Symmetry, beauty and evolution. *Nature, 372,* 169-172.

Enquist, M. & Arak, A. (1998). Neural representation and the evolution of signal form. In R. Dukas (Ed.), *Cognitive ecology* (pp. 21-87). Chicago: University of Chicago Press.

Enquist, M. & Ghirlanda, S. (1998). The secrets of faces. *Nature, 394,* 826-827.

Etcoff, N. L. (1999). *Survival of the prettiest: the science of beauty.* New York: Doubleday.

Fagen, R. (1974). Selective and evolutionary aspects of animal play. *American Naturalist, 108,* 850-858.

Falconer, D. S. (1989). *Introduction to quantitative genetics.* 3rd ed. Harlow: Longman Scientific, and New York: Wiley.

Faltings, B. (1997). Qualitative models as indices for memory-based prediction. *IEEE Expert, 12,* 47-53.

Feist, J. (Ed.). (2001). Special issue on aesthetics. *Bulletin of Psychology and the Arts 2*(1).

Fielding, N. G. & Lee, R. M. (1991). *Using computers in qualitative research.* London: Sage Publications.

Fisher, P. (1999). Wonder, the rainbow, and the aesthetics of rare experiences. New York: Cambridge.

Fitzgerald, R. J. (1971). Divergence of the Kalman filter. *IEEE Transactions on Automatic Control, 16,* 736-747.

Flack, V. F. & Chang, P. C. (1987). Frequency of selecting noise variables in subset regression analysis: a simulation study. *American Statistician, 41,* 84-86.

Fouché, P. & Kuipers, B. (1990). *An assessment of current qualitative simulation techniques.* Report AI90-140, Artificial Intelligence Laboratory, University of Texas at Austin.

Fowler, C. (1988). Population dynamics as related to rate of increase per generation. *Evolutionary Ecology, 2,* 197-205.

Gadgil, M. D. (1972). Male dimorphism as a consequence of sexual selection. *American Naturalist, 106,* 574-580.

Geist, V. (1978). *Life strategies, human evolution, environmental design: To-*

wards a biological theory of health. New York: Springer-Verlag.

Gerlach, S. C. & Murray, M.S. (2001). *People and wildlife in northern North America: Essays in honor of R. Dale Guthrie.* Oxford: Archaeopress.

Ghiselin, M. T. (1969). *The triumph of the Darwinian method.* Berkeley: University of California Press.

Ghiselin, M. T. (1974). *The economy of nature and the evolution of sex.* Berkeley: University of California Press.

Gombrich, E. H. (1961). *Art and illusion.* 2nd ed. Princeton: Princeton University Press.

Goodwin, B. (1994). *How the leopard changed its spots.* Princeton: Princeton University Press.

Götmark, F. & Ahlström, M. (1997). Parental preference for red mouth of chicks in a songbird. *Proceedings of the Royal Society of London B, 264,* 959-962.

Gould, J. L., Elliott, S. L, Masters, C. M. & Mukerji, J. (1999). Female preferences in a fish genus without female mate choice. *Current Biology, 9,* 497-500.

Gould, S. J. (1991). The great seal principle. *Natural History, 100* (March), 4-12.

Graham, R.W. et al. (20 co-authors). (1996). Spatial response of mammals to late quaternary environmental fluctuations. *Science, 272,*1601-1606.

Guthrie, R. D. (1995). Mammalian evolution in response to the Pleistocene-Holocene transition and the break-up of the mammoth steppe: Two case studies. *Acta Zoologica Cracovensia, 38,* 139-154.

Guthrie, R. D. (2001). Origin and causes of the mammoth steppe: A story of cloud cover, woolly mammoth tooth pits, and inside-out Beringia. *Quaternary Science Reviews, 20,* 549-574.

Hampl, P. (1999). *A romantic education.* New York: Norton.

Handelman, D. (1992). Passages to play: Paradox and process. *Play and Culture, 5,* 1-19.

Handelman, D. & Shulman, D. (1997). *God inside out: Siva's game of dice.* New York: Oxford University Press.

Harvey, A. C. (1990). *Forecasting, structural time series models, and the Kalman filter.* New York: Cambridge University Press.

Haykin, S. S. (2002). *Kalman filtering and neural networks.* New York: Wiley.

Hearne, V. (1986). *Adam's task: Calling animals by name.* New York: Knopf.

Hein, N. (1987). Lila. In M. Eliade (Gen. Ed.), *Encyclopedia of religion, vol. 8.* New York: Macmillan.

Heinemann, J. A. (1993). Bateson and peacocks' tails. *Nature, 363*:308.

Hinde, R. A. (1985). Was 'The expression of the emotions' a misleading phrase? *Animal Behaviour, 33,* 985-922.

Johnsgard, P. A. (1994). *Arena birds: Sexual selection and behavior.* Washington: Smithsonian Institution Press.

Johnson, K. & Coates, S.L. (2001). *Nabokov's butterflies.* New York: McGraw-Hill.

Josipovici, G. (1996). *Touch.* New Haven: Yale University Press.

Kamo, M. & Iwasa, Y. (2000). Evolution of preference for consonances as a by-product. *Evolutionary Ecology Research, 2,* 375-383.
Kamo, M., Kubo, T. & Iwasa, Y. (1998). Neural network for female mate preference, trained by a genetic algorithm. *Philosophical Transactions of the Royal Society B, 353,* 399-406.
Kirkpatrick, M. (1982). Sexual selection and the evolution of female choice. *Evolution, 36,* 1-12.
Klopfer, P. H. (1970). Sensory physiology and esthetics. *American Scientist, 58,* 399-403.
Kortmulder, K. (1998). *Play and evolution.* Utrecht: International Books.
Krebs, J. R. (1977). The significance of song repertoires: The Beau Geste hypothesis. *Animal Behaviour, 25,* 475-478.
Kubler, G. (1962). *The shape of time.* New Haven: Yale University Press.
Lakoff, G. & Johnson, M. (1980). *Metaphors we live by.* Chicago: University of Chicago Press.
Lakoff, G. & Johnson, M. (1999). *Philosophy in the flesh.* New York: Basic Books.
Lambert, C. (1999). The stirring of sleeping beauty. *Harvard Magazine, 102*(1), 46-53.
Lande, R. (1981). Models of speciation by sexual selection on polygenic traits. *Proceedings of the National Academy of Sciences USA, 78,* 3721-3725.
Lande, R. (1982). Rapid origin of sexual isolation and character divergence in a cline. *Evolution, 36,* 213-223.
Levi-Strauss, C. (1992). *Tristes tropiques.* Trans. J. and D. Weightman. New York: Penguin.
Lewontin, R. C. (1977). Sociobiology: A caricature of Darwinism. In P. Asquith & F. Suppe (Eds.), *PSA 1976, vol. 2.* (pp. 22-31). East Lansing, MI: Philosophy of Science Association.
Lewontin, R. C. (1974). Annotation. The analysis of variance and the analysis of causes. *American Journal of Human Genetics, 26,* 400-411.
Lumsden, C. (1991). Aesthetics. In M. Maxwell (Ed.), *The sociobiological imagination* (pp. 253-268). Albany, NY: State University of New York Press.
Lumsden, C. (2001). Fit to be eyed: Genes, culture and creative minds. *Bulletin of Psychology and the Arts, 2*(1), 16-20.
Manning, J. T. Trivers, R. L., Singh, D., & Thronhill, R. (1999). The mystery of female beauty. *Nature, 399,* 214-215.
Masson, J. M. & McCarthy, S. (1995). *When elephants weep: The emotional lives of animals.* New York: Delacorte Press.
Moller, A. P. (1994). *Sexual selection and the barn swallow.* New York: Oxford University Press.
Moller, A. P. & Sorci, G. (1998). Insect preference for symmetrical artificial flowers. *Oecologia, 114,* 37-42.
Moore, G. E. (1903). *Principia Ethica.* Cambridge: Cambridge University Press.
Morris, D. (1962). *The biology of art.* New York: Knopf.
Morris, M. R. (1998). Female preference for trait symmetry in addition to trait

size in swordtail fish. *Proceedings of the Royal Society of London Biological Sciences, 265,* 907-912.

Morris, M. R. & Casey, K. (1998). Female swordtail fish prefer symmetrical sexual signal. *Animal Behaviour, 55,* 33-39.

NASA. (1993). *Qualitative model-based diagnostics for rocket systems.* United States National Aeronautics and Space Administration report N 93-28052. Springfield, VA: National Technical Information Service.

Nabokov, V. (1962). *Pale fire.* New York: Putnam.

Nabokov, V. (1989). *Speak, memory.* New York: Vintage International.

Nabokov, V. (2000). *Nabokov's butterflies.* Boston: Beacon Press.

Nelson, K. (1973). Does the holistic study of behavior have a future? In G. Bateson & P. H. Klopfer (Eds.), *Perspectives in ethology, Vol. I* (pp. 281-328). London: Plenum Press.

Nelson, M. M., & Illingworth, W. T. (1990). *A practical guide to neural nets.* Reading, MA: Addison-Wesley.

Neter, G., Kutner, M. H., Wasserman, W., & Nachtsheim, C. (1996) *Applied linear statistical models.* 4th ed. New York: McGraw-Hill.

Orians, G. (1986). An ecological and evolutionary approach to landscape aesthetics. In E. C. Penning-Rowsell and D. Lowenthal (Eds.), *Landscape meanings and values* (pp. 3-25). London: Allen and Unwin.

Pielou, E. C. (1988). *The world of northern evergreens.* Ithaca, NY: Comstock Publishing Associates at Cornell University Press.

Pielou, E. C. (1991). *After the ice age.* Chicago: University of Chicago Press.

Pielou, E. C. (1994). *A naturalist's guide to the Arctic.* Chicago, IL: University of Chicago Press.

Potegal, M. & Einon, D. (1989). Aggressive behavior in rats deprived of play-fighting experience as juveniles. *Developmental Psychobiology, 22,* 159-172.

Praz, M. (1967). *Mnemosyne: The parallel between literature and the visual arts.* Princeton: Princeton University Press.

Proust, M. (1989). *À la recherche du temps perdu.* Vol. IV. J.-Y. Tadié (Ed.). Paris: Gallimard.

Reece, S. (1998). *Qualitative model-based multi-sensor data fusion: The qualitative Kalman filter.* D. Phil. thesis, Oxford University.

Rice, E. L. & Chapman, A. R. O. (1985). A numerical taxonomic study of Fucus distichus (Phaeophyta). *Journal of the Marine Biological Association of the United Kingdom, 65,* 433-459.

Richardson, A. (2001). Annotated bibliography on literature, cognition & the brain. URL http://www2.bc.edu/~richarad/lcb/bib/annot.html.

Ripley, B. D. (1996). Pattern recognition and neural networks. Cambridge: Cambridge University Press.

Roberts, J. M., Mills, D. J., Charnley, D. & Harris, C. J. (1996). Improved Kalman filter initialisation using neurofuzzy estimation. *University of Southampton Department of Electronics and Computer Science Research Journal,* on line.

Rosenthal, G. G. & Evans, C. S. (1998). Female preference for swords in *Xiphophorus helleri* reflects a bias for large apparent size. *Proceedings of the National Academy of Sciences USA, 95,* 4431-4436.

Ryan, M. J. (1991). Sexual selection, sensory systems, and sensory exploitation. *Oxford Surveys in Evolutionary Biology, 7,* 157-195.

Ryan, M. J. (1998). Sexual selection, receiver biases, and the evolution of sex differences. *Science, 281,* 1999-2003.

Sacks, O. (1998). *A leg to stand on.* New York: Touchstone Books, Simon & Schuster.

Scarry, E. (1999). *On beauty and being just.* Princeton: Princeton University Press.

Schmalhausen, I. I. (1949). *Factors of evolution.* Philadelphia, PA: Blakiston.

Sewell, E. (1952). *The field of nonsense.* London: Chatto & Windus.

Sheets-Johnstone, M. (1990). *The roots of thinking.* Philadelphia, PA: Temple University Press.

Sheets-Johnstone, M. (1999). *The primacy of movement.* Philadelphia, PA: John Benjamins Publishing Co.

Southwood, T. R. E. & Kennedy, C. E. J. (1983). Trees as islands. *Oikos, 41,* 359-371.

Sparshott, F. (1982). *The theory of the arts.* Princeton: Princeton University Press.

Sparshott, F. (1995). *A measured pace.* Toronto and Buffalo: University of Toronto Press.

Sparshott, F. (1998). *The future of aesthetics: The 1996 Ryle lectures.* Toronto and Buffalo: University of Toronto Press.

Spinka, M., Newberry, R. C. & Bekoff, M. (2001). Mammalian play: Training for the unexpected. *Quarterly Review of Biology,* 76, 141- 168.

Steadman, P. (1979). *The evolution of designs.* Cambridge, MA: Cambridge University Press.

Sutton-Smith, B. (1997). *The ambiguity of play.* Cambridge, MA: Harvard University Press.

Sutton-Smith, B. (2003). Play as a paradox of emotions. In D. Lytle (Ed.), *Play and educational theory and practice.* Play & Culture Studies (Vol. 5, pp. 3-17). Westport, CT: Praeger.

Suzuki, J. (1990). *Plant control expert system coping with unforeseen events - model-based reasoning using fuzzy qualitative reasoning.* Tokyo: Institute for New Generation Computer Technology.

Symons, D. (1995). Beauty is in the adaptations of the beholder: The evolutionary psychology of human female sexual attractiveness. In P. R. Abramson & S.D. Pinkerton (Eds.), *Sexual nature, sexual culture* (pp. 80-118). Chicago: University of Chicago Press.

Thornhill, R. (1998). Darwinian aesthetics. In C. Crawford and D. Krebs (Eds.), *Handbook of evolutionary psychology* (pp. 543-572). Mahwah, NJ: Erlbaum.

Tovée, M. J. & Cornelissen, P. L. (1999). The mystery of female beauty. *Nature, 399,* 215-216.

Trivers, R. (1985). *Social evolution.* Menlo Park, CA: Benjamin/Cummings.

Tuan, Y.-F. (1974). *Topophilia.* Englewood Cliffs, NJ: Prentice-Hall.

Tuan, Y.-F. (1993). *Passing strange and wonderful: Aesthetics, nature, and culture*. Washington, DC: Island Press.

Valéry, P. (1964). *Aesthetics*. Tran. Ralph Manheim. New York: Pantheon Books.

Waddington, C. H. (1957). *The strategy of the genes*. London: Allen and Unwin.

Washburn, S. L., & Hamburg, D.A. (1965). The study of primate behavior. In I. DeVore (Ed.), *Primate behavior* (pp. 1-13). New York: Holt, Rinehart, & Winston.

Wells, C. (1996). *The Kalman filter in finance*. Dordrecht and Boston: Kluwer Academic Publishers.

Wilson, E. O. (1998). *Consilience*. New York: Knopf.

Yu, D. W. & Shepard, G. H. (1998). Is beauty in the eye of the beholder? *Nature, 396,* 321-322.

Yu, D. W. & Shepard, G. H. (1999). The mystery of female beauty: Reply. *Nature, 399,* 216.

Zarchan, P. & Musoff, H. (2000). *Fundamentals of Kalman filtering: A practical approach*. Reston, VA: American Institute of Aeronautics and Astronautics.

Chapter 2

Videotape Entertainment May Facilitate Recovery for Monkeys in a Clinical Setting

Peggy O'Neill-Wagner

INTRODUCTION

Results of research studies investigating the effects of video stimulation as enrichment for monkeys have been few and mixed. Findings of Platt and Novak
(1997) clearly indicate that two solitary adult male rhesus monkeys viewed
videotapes with an audio component and that the exposure promoted their well
being by increasing their activity levels. The two males preferred videotapes
containing unfamiliar monkeys and humans, although they eventually habituated
to repeated programs. In contrast, a study done by Schapiro and Bloomsmith
(1995) indicates that when singly housed yearling rhesus, housed outdoors, were
exposed to videotaped programs of monkeys with no audio component they
showed almost no interest. In support of Platt's findings, an informal descriptive
report from the Duke University Medical Center (Heath, 1987) indicates that
even a black and white TV, offering visual and auditory stimulation, seemed to
reduce abnormal behavior, while increasing overall activity and vocal behavior
of singly caged rhesus monkeys and baboons. Additionally, to add support to the
plausibility of reducing boredom with videotapes in a clinical setting, an animal
learning study by Swartz and Rosenblum (1980) found that juvenile bonnet macaques responded at high-sustained levels in an operant task for hours to obtain
a videotape reward presenting color images of monkeys. Their research concluded that the images were socially stimulating and valuable to the young monkeys viewing them. With this in mind one might consider offering videotape
programming to hospitalized monkeys.

According to Humphrey's "interest" and "pleasure" theory (1973) for determining rhesus monkeys' visual preferences, the contrasting outcomes for the
Platt and Schapiro studies probably differed for good reason. More initial 'interest' was stimulated by the video and audio combination, even though it waned

after repeated showings. But 'pleasure' derived from the content wasn't sufficient to keep them watching repeatedly. For the yearling monkeys it is likely that there was neither 'interest' nor 'pleasure' derived from the repetitious silent videotapes because the young monkeys were quarantined outdoors where they could observe real primates interacting. The outdoor environment offered more complex sensory stimulation than the silent tapes and was thereby more likely to provide 'interest' and/or 'pleasure.' The only factor the two studies had in common was repetition. The first was repetitious for program, the second for content. Informal observations made by Heath (1987) do not indicate if either type of repetition occurred.

For successful application of videotape enrichment at our research facility it is important that the enrichment appeal to: (1) male and female rhesus monkeys, (2) of all ages, (3) coming from indoor and outdoor habitats, (4) while undergoing medical treatment, and (5) experiencing social separation from their social groups. It is not important that the animals remain attentive to the tapes for months or years, however. It seemed worthwhile to test the conflicting research reports, as there were no data suggesting that the animal's well being could be at risk from exposure to videotape. And, if a theory promoted by the research of Butler (1961), and Butler and Woolpy (1963) was accurate, then the "continuously changing stimuli" offered by the videotape format would "maintain the monkeys interest longer." By offering varied videotape content without repeated tape showings perhaps 'Interest' and/or 'Pleasure' could be maximized and thereby reduce or eliminate the opportunity for removal of stitches and bandages during isolation and recovery.

METHODS

Subjects

Ten rhesus monkeys (*Macaca mulatta*) undergoing treatment in the research center's veterinary clinic served as research subjects during the course of their recoveries. Their ages, gender, housing arrangements (outside the clinic) and rearing experience are all listed in Table 2.1. Each animal was confined to a single cage while undergoing treatment, lasting between three and fourteen days, depending upon the individualized medical protocol. In seven cases our subjects were being treated for fight wounds, in two cases for diarrhea, and one was recovering from surgery. Most animals came in from social groups engaged in the typical rivalries and social stresses of the breeding season. In many cases animals were alone in the treatment room for all or part of their recovery period.

TABLE 2.1
Monkey Subjects

	AGE	REARING	HOUSING
ADULT MALES:			
MAYNARD	9	MOTHER	SINGLE/GROUP (BREEDER)
NORMAN	12	MOTHER	SINGLE/GROUP (BREEDER)
HERSHEY	15	MOTHER	TROOP/FIELD
ADULT FEMALE:			
EMMA	20	MOTHER	TROOP/FIELD
ADOLESCENT FEMALES:			
ZIPPER	3.5	MOTHER	TROOP/FIELD
LUCY	3.5	MOTHER	TROOP/FIELD
JUVENILE MALES:			
BEEPER	2.5	PEER	GROUP
MILES	1.5	MOTHER	TROOP/FIELD
JUVENILE FEMALES:			
JUANITA	2.5	PEER	GROUP
PEROT	2.5	PEER	GROUP

Equipment

All treatment cages were stainless steel primate holding cages and each was a size determined to be compliant with the USDA animal size and weight standards. Each cage was complete with a perch and at least one manipulative object and/or mirror. The dietary and fluid intake offered to each animal varied according to the specified health care protocol. Videotape apparatus consisted of a 13-inch Panasonic Combination TV-Video Cassette Recorder on a cart. A Panasonic VHS Camcorder and tripod were used to document the animals' response during and after exposure to videotape. Videotape selections are listed in Table 2.2. The longer the monkeys spent in recovery, the more videotapes they had the opportunity to view. Each videotape was played only once to each animal.

TABLE 2.2
Featured Videotapes

Primate Documentaries

   World of Primates
   Mandrills of Gabon

Macaques

   NIH Field Station Breeding

Animation

   Fantasia

People and Animals

   Junglebook

Soap Opera

   The Young and the Restless

Data Collection

The cart and TV-VCR unit were kept in the treatment room to develop animal familiarity and comfort with the equipment. Following all daily animal treatments the "entertainment" cart was rolled toward the animal cage. By showing selected feature programs after daily treatment procedures the animal would not subject to distraction or stressful interruption. While tapes were showing, the monitor was placed less than 1 m from the cage (unless the animal showed signs of discomfort—one animal was only comfortable with it farther away). At this period of the day the adjacent room and hallway areas were generally quiet. Videotapes were shown to the monkeys on weekdays and on weekends. Each tape played for between 90 and 105 minutes. The Camcorder collected behavioral data for a full two-hour period. In this way the monkey's behavior could also be observed when there was no taped program showing. Following data collection tapes were removed from the camcorder and taken to the research lab. They were played back and scored for six visual orientation categories (Table 2.3) during eight 5-minute segments throughout each tape. Five-minute segments were balanced for early, midway, and late program time.

TABLE 2.3
Visual Orientation Categories

HIGH ATTENTIVENESS

Avid:
Subject remains visually oriented toward screen during most of the program and may engage in other activity while looking toward the monitor.

Attentive:
Subject visually orients back and forth to screen and away from screen for approximately equal intervals of time during the program, and maintains some degree of eye contact.

MODERATE ATTENTIVENESS

Selective Attentive:
Subject is primarily engaged in other activity, but stops activity and looks toward monitor, as if interrupted. Time blocks are variable while attending/not attending visually to screen. May include some glancing.

Combination Attentive:
Subjects' primary visual orientation can best be described by more than one category.

LOW ATTENTIVENESS

Routinely Attentive:
Subject visually orients toward the TV screen as if surveying or scanning: Glances toward the TV screen regularly. Does not maintain eye contact. May be incorporated into stereotypic movements.

No Visual Attentiveness:
Subject may engage in sleep, passive, and/or TV monitor avoidance for most of the presentation.

RESULTS AND DISCUSSION

The observations summarized in Table 2.4 show that when introduced to videotapes rhesus monkeys, for the most part, display a degree of visual attentiveness toward videotape programming. All subjects fluctuated in their attentiveness as program content changed. Gender and rearing history appear to have little relationship to overall interest in videotapes. Very young monkeys maintain much shorter periods of attentiveness as they engage in more playing, locomotion, and exploring of their cages. Two adults, one male (Maynard) and one female (Emma) stood out as the most consistently 'Avid' viewers. Maynard showed no interest in the soap opera, but avidly watched the other programs for hours on end. One juvenile female (Juanita) was always scored as a 'Moderate' or highly attentive viewer, regardless of the program. Lucy (adolescent female) wasn't exposed to the soap opera, but she showed a high level of interest in three of the four programs. The only one that she showed 'Low' interest toward received 'High' attentiveness scores from two other females, a juvenile and an adult. Two males, one juvenile (Miles) and one adult (Hershey), showed little interest in watching videotapes. But, data notes indicate, "Hershey appeared to be listening to the audio on the tapes". Other animals had visual attentiveness scores that varied from 'High' to 'Low' (Lucy and Beeper) relative to tape content.

TABLE 2.4
Individual Ratings for Visual Orientation

Video	Content	Attentiveness		
		High	Moderate	Low
Junglebook	People/animal	Maynard Emma Juanita	Perot	Hershey Zipper Lucy Miles
Macaques	Breeding	Maynard Emma Lucy Juanita Beeper	Hershey Zipper	Norman Perot Miles

Continued on next page

Table 2.4 - *Continued*

Fantasia	Animation	Maynard Lucy	Juanita	Hershey Zipper Perot Miles Beeper
The Young & Restless	Human Activity		Emma Juanita	Maynard Zipper
Primate Documentary	Monkey/Apes	Maynard Emma Lucy	Zipper Juanita Perot	Hershey Miles Beeper

Though individual variation seems an appropriate way to describe the findings, there is a factor that wasn't originally considered which may help to understand the results. Four animals that were ranked as 'Highly Attentive' viewers during at least 50 percent of the videotapes are low ranking in their social groups. While in their social groups they had engaged in regular surveillance of their surroundings to avoid being targeted during social conflicts. By contrast, the three highest-ranking monkeys in this study, two males and one female, were consistently rated as 'Low' in attentiveness to videotape, regardless of program content. Not only does the need to be watchful of others differ between animals due to rank, but also the ability to control the environment is also profoundly different. Low ranking animals probably have adjusted to having less control in their lives, and may actually enjoy watching videotaped interactions of others from the protected cage. Higher-ranking animals, on the other hand, may be frustrated by their inability to intervene and control the outcome of the activity they observe on tape. Highest levels of emotive/excited activity involving cage shaking, dominance displays and aggressive gestures toward other animals in the treatment room and toward the television monitor were initiated by high ranking animals during videotaped segments of nonhuman primate breeding and social interaction.

The Field Breeding Season videotape received 'High' and 'Moderate' levels of attentiveness from 70% of the monkeys. Five of the treatment room subjects came from that field troop and thus were watching familiar troop members. Four of the five animals originating from the field environment watched the tape. In fact, Emma and Lucy watched avidly. To the other five subjects the monkeys in the breeding video were strangers. Three of the five monkeys watched the strangers with 'Avid' levels of attentiveness. It is not surprising that animals removed from their social groups during the breeding season when hormone levels are in flux showed high levels of attentiveness during tape segments showing sex skin coloration, breeding, postural presentation, dominance, and aggression. The Documentary tapes showing other primates rated second with 66.6% of the subjects exhibiting 'High' or 'Moderate' levels of attentiveness. Both Jungle Book and the Soap Opera had 50% of the monkeys watching, but 37.5% were watching Jungle Book with 'High' levels of attentiveness, whereas, no animals watching the Soap Opera were scored as 'High' in attentiveness. Fantasia, a fantasy cartoon style tape with lots of classical music for audio background to the colorful and creative visuals, had only 37.5% of the monkeys watching with 'High' and 'Moderate' levels of attentiveness.

One unanticipated observation was recorded numerous times on the data score sheets. When monkeys on the videotapes were eating, the animals viewing the tapes tended to start eating biscuits and other food available in their cages. There is no way to know if there was an actual correlation between the two events, but it might serve as an interesting topic for a controlled study. If separation anxiety leading to food refusal during clinical treatment can be resolved by showing stimulating videotapes, this could be a valuable therapeutic option.

Due to the uncontrolled nature of a clinical setting these findings are descriptive but not scientifically conclusive. There was no control group. There was variation in the amount of videotape exposure each subject received. The treatment, feeding, handling, and sedation of animals varied during the course of these observations, according to the individualized health care protocol. Some patients were alone in the treatment room for all or part of their stay. Patients were admitted and released as conditions warranted it necessary. In spite of these obvious confounds, the physical setting remained the same throughout the observation period and the procedures of video presentation remained constant.

CONCLUSIONS

Monkeys do maintain some degree of visual and auditory attentiveness toward videotape programming.

Videotape viewing interest and activity by rhesus monkeys is subject to individual variation, and can serve as a valuable enrichment option.

Noises and activities ongoing in the hallway, adjacent areas and inside the treatment room temporarily distract monkeys from watching videotapes.

Videotape viewing by monkeys was a pleasant and sometimes entertaining experience for the animal care staff.

Videotape presentation is a safe and convenient way to offer visually stimulating enrichment to monkeys in a clinical setting and may provide diversion from pulling at sutures and removing bandaging.

Important components of videotape enrichment success are selection of appropriate program content, volume, monitor placement and variety.

ACKNOWLEDGEMENTS

This research has been supported by intramural funds of the National Institute for Child Health and Human Development. Acknowledgement is extended to Dr. Rosemary Bolig, Child Associates, Washington, D.C. for her valued personal support and advice and to Veterinarian Dr. Lynn Walker for her professional expertise, patience, and staff support throughout this study. I would also like to thank two animal care technicians, Mr. Edward George and Mr. Lynn Roberts for their patience, daily feedback, and initial observations, which were the stimulus for this study.

REFERENCES

Butler, R. A. (1961). The responsiveness of rhesus monkeys to motion pictures, *Journal of Genetic Psychology, 98*, 239-245.

Butler, R. A. & Woolpy, J. H. (1963). Visual attention in the rhesus monkey, *Journal of Comparative Physiology, 56*, 324-38.

Heath, S. J. (1987). Behavioral enrichment for primates: What are the options? *Scientist's Center for Animal Welfare Newsletter, 9*(1), 11-12.

Humphrey, N. K. (1972). 'Interest' and 'pleasure:' Two determinants of a monkey's visual preferences. *Perception, 1*, 395-416.

Meunier, L. D., Duktig, J. T. & Landi, M. S. (1989). Modification of stereotypic behavior in rhesus monkeys using videotapes, puzzle feeders, and foraging boxes. *Laboratory Animal Science, 39*(5), 479.

Platt, D. M. & Novak, M. A. (1997). Video stimulation as enrichment for captive rhesus monkeys (*Macaca mulatta*). *Applied Animal Behavior Science, 52*, 139-155.

Schapiro, S. J. & Bloomsmith, M. A. (1995). Behavioral effects of enrichment on singly housed, yearling rhesus monkeys: An analysis including three enrichment conditions and a control group. *American Journal of Primatology, 35*, 89-101.

Swartz, K. B. & Rosenblum, L. A. (1980). Operant responding by bonnet macaques for color videotape recordings of social stimuli, *Animal Learning and Behavior, 8*, 311-321.

Chapter 3

A Qualitative Approach to Boys Rough and Tumble Play: There is More Than Meets the Eye

Thomas L. Reed

Play is something that is universal but yet is not universally defined. The difficulty of defining play lies in that it can mean different things to different players. Play theorists contribute to the difficulty of defining play with followers of Vygotsky, postulating that play is influenced by culture (Cole et al., 1978) and Piagetians believe that play follows a predictable developmental path (Piaget & Inhelder, 1969). There are, however, some agreed upon commonalities about play. One of the commonalities in defining play that is accepted by most researchers is the assertion by Csikszentmihalyi (1975) that play is fun and something we feel not something we do. That is to say that play is something that is felt internally as well as the more commonly accepted external expression of feelings. Legitimate forms of play will not result in psychological or physical victimization of children (MacDonald, 1992). Truly, the player is the individual most likely to accurately determine if what he or she is engaged in is indeed a playful experience. There is one category of play that is often confused with aggressive behavior. Rough and tumble (R&T) play is a label for a type of play that when observed may appear to be aggressive and during which physical harm may occur.

DEFINING ROUGH AND TUMBLE PLAY

Rough and tumble (R&T) play is a type of play that was brought to popular notice by Harlow's (1962) research with Rhesus monkeys. Harlow described the basic tenets of R&T play as running, chasing and fleeing, wrestling, open hand beating, falling, and play fighting. N. Blurton Jones (1976) found that the same basic characteristics and strategies of R&T play discussed by Harlow also applied to children playing in a nursery school. However, R&T play is observed in both genders in the Rhesus monkeys, in humans it is predominantly, although not exclusively, observed in males (Blurton Jones, 1976; Fagot, 1984; Pellegrini and Perlmutter, 1988; Pellegrini, 1988 & 1995; and MacDonald, 1992). It has been noted that the behaviors involved in R&T play can be confused with fighting or slapping (Blurton Jones, 1976) and indeed one definition of R&T play is "play fighting." When children engaged in R&T play are admonished to stop fighting they often claim that they were "just playing." Furthermore, Pellegrini (1989) has found that aggressive play amounts to less than 5 percent of all play on the playground. Rough and tumble behaviors have become institutionalized in time-honored games such as "King of the Mountain" where the object is to playfully wrestle the person off the top of the hill and "Red Rover" where children break through a human chain in a fun-loving way. Even the classic, "I'm Gonna Get You" often evokes responses which outwardly appear aggressive and closely resemble R&T behavior (Sutton-Smith, 1972).

In addition to being fun for the players, there are benefits to children permitted to participate in R&T play. Engagement in R&T play has been positively correlated with social problem solving ability and academic achievement (Pellegrini, 1994). Research conducted by Pellegrini and Perlmutter (1988) established that some of the ways in which social competence is developed are alternating role taking, give and take, deciding who is to follow and who is to lead, and initiating social dominance.

Donaldson (1976) suggests that children express feelings for one another during R&T play. The skills acquired in R&T play can be used by adolescents to contribute to their competence in organized sports (Sutton-Smith, 1992). Play theorists have attempted to explain how children are able to know the difference between R&T play and aggression. There is a form of communication called metacommunication that appears to bridge the gap between the outward appearances of aggressive play and care expressed by boys (Donaldson, 1976). The uninformed observer of R&T play may see tripping, pushing, or hitting and other aggressive behavior and fear that this behavior will lead to fighting. While playing with his subjects, Donaldson discovered instead of aggression, children were seeing hugging, loving, compassion, embracing, mutual sharing, and caring for one another. Donaldson further theorizes that the R&T player is saying, "I trust you to push me, trip me, fall on me, and if I get hurt you will care for me" (p. 239). In addition, Donaldson says that R&T play is non-competitive and a player's skill level is relatively unimportant (1976).

Pre-adolescent boys, according to Gilligan (1997), are very affectionate and enjoy physical and emotional closeness with their parents. Boys like to be hugged, kissed, and freely participate in overt expressions of affection. Gilligan asserts that, boys about age three begin a gradual withholding of the outward expression of feelings. For example, a boy who once enjoyed an affectionate nickname now becomes agitated particularly when it is used around his friends. In the comic strip, *Dennis the Menace* by Hank Ketchum, Dennis is being hugged by his mother as he is getting ready to get on the school bus. Several of his friends are waiting for him and Dennis says to his mother, "Don't call me

Honey. Call me Slugger." This illustrates perfectly Gilligan's point that at an early age boys learn to withhold their feelings.

Homophobia among adults makes it difficult for children, especially boys, to express feelings of affection to same sex members. At an early age many boys realize that an innocent gesture of affection such as holding another boy's hand may result in ridicule or shame from family and friends. Donaldson (1976) proposes that one way in which boys express care for one another is through R&T play which can be seen as early as eighteen months (Blurton Jones, 1976).

Since many R&T behaviors resemble aggressive actions, some researchers consider R&T play to be a form of antisocial behavior (Bredekamp, 1992; Ladd, 1983) which they feel should be discouraged. Bredekamp (1992) calls for intervention when children "get carried away" with chasing or wrestling (p. 74). Yet, chasing and wrestling are two salient characteristics of R&T play and to the uninformed observer may appear to what Bredekamp identifies as "over stimulated children" (p. 74). There has been modification of this position in the 1997 edition of NAEYC's DAP guidelines indicating that R&T is acceptable for preschool aged children (Bredecamp & Copple, 1997).

The purpose of the ethnographic study was to examine and describe the way in which preadolescent boys in a natural setting express care for one another through R&T play. Caring, like playing, is a universal concept but not easily defined. Noddings (1990) calls caring the most significant form of human response. Caring, compassion, and connection are important for all people to understand and value (Noddings, 1990). Donaldson (1976) theorizes that care is expressed through R&T play. This research study is designed to test this theory and to answer the following questions: (1) How do boys set up the R&T experience? and (2) How do they express care for one another through a set of behaviors that appear to be aggressive in nature?

METHODOLOGY

Setting and Activity Schedule

The study was conducted at a Youth Center in a military installation Youth Center located in the southeastern United States. The primary purpose for the Youth Center is to provide before and after school programs for school aged children (grades 1-7). The facility is staffed by recreation specialists who receive ongoing training in child development. The Youth Center is accredited by the Department of Defense and ascribes to the Wellesley School Age Program educational philosophy.

The afternoon schedule for the Youth Center begins daily at 2:30 p.m. with approximately thirty minutes of free play allowing the students to engage in a variety of activity centers including art, reading, housekeeping, dramatic play, and outdoor play (or in the gymnasium during inclement weather). This is followed by a snack and students are then permitted to participate in structured activities or to continue in free play. Students are to be picked up by 5:30 each day. Since students do not play outdoors in the morning, only the afternoon period was used for this study.

Procedures

The study was ethnographic and this perspective was used to analyze the data. The observations and subsequent interviews of R&T players were recorded on videotape by the researcher. The data were collected over a five-day period. The researcher videotaped children playing from approximately 2:30 to 5:30 or whenever R&T play occurred during the data collection period. From a total of 15 hours of observing students on the playground, the researcher was able to capture a total of three hours of play that met the preset qualifications of R&T play (Table 1). Credibility and internal validity in this study were strengthened by persistent observation, prolonged engagement in R&T play, member checking, triangulation, and peer debriefing as described by Lincoln and Guba (1985). Member checking was carried out with three early childhood professionals (adjudicators) who viewed the videotapes. The adjudicators (two of whom hold Ph.D.s and one doctoral candidate) achieved an inter-rater reliability of 97.4%.

Subjects

The student population of the Youth Center is diverse with White, Black, Asian, and Hispanic children in attendance. The selection of children for inclusion in the study was based on their engagement in R&T play. Table 3.1 lists the characteristics (Blurton Jones, 1976). Seven boys whose ages ranged from six through nine (mean age=7.42) met the criteria previously listed in Table 3.1. No girls chose to engage in R&T play. The participants were interviewed individually and as a group while they were observing videotapes of selected R&T play episodes of themselves and others.

Table 3.1
Rough and Tumble Play Characteristics (Blurton Jones, 1976)

Chase	Playfighting
Flee	Poke
Hit at	Pounce
Hold	Push
Kick	Tease
Pile on	Wrestle

The seven boys that met the criteria during the data collection participated in an R&T game called Smear the Queer (making their selection one of convenience). The racial makeup of the participants (4 white and 3 black) selected for this study was incidental (Table 3.2) and dependent upon the participants who played Smear the Queer. The game of Smear the Queer created a special closeness among the boys and the game itself was a representative sample of R&T play as listed in Table 3.1. It should be noted that despite the repugnant nature of the name "Smear the Queer" (Smear), it was the players who coined the term. The name Smear was never used in any other context by the players other than the identification of the game they were playing. There were a few others who entered into the game of Smear; however, they did not stay long and were not encouraged (nor discouraged) to return. The seven participants played Smear together everyday, looked for one another to get the game started, and otherwise considered themselves to be friends.

Table 3.2
Age and Race of Participants

Age	Race
6	Black
7	White
7.5	Black
7.5	Black
8	White
8	White
9	White

The game of Smear bears a striking resemblance to American football in that a ball is launched by a player and chased after by any number of players all of whom are eligible to capture the ball. Once a person retrieves the ball, he attempts to return to a pre-selected spot. At this point he may be tackled, knocked over, tripped, or gang tackled. At the point of tackle, the ball is again launched and the process repeats itself.

The participants, despite playing the same game together everyday and they are as follows (their names have been changed): Kevin (age 9), Brian (age 8), Jake (age 8) identical twins Perry and Patrick (ages 7.5), Zach (age 8), and Linden (age 6).

All play that resembled R&T play was recorded. Initially, students "hammed it up" for the camera but quickly fell into their normal play patterns. Smear was played in a specific area of the playground, while other episodes of R&T occurred at random. The researcher was able to be within a few feet of most of the videotaped R&T play.

The first step of the data analysis involved reviewing the videotaped play searching for and identifying isolated examples of R&T play, (Table 3.1), as well as identifying R&T play behaviors found in the game of Smear. In addition, the videotape was again reviewed, searching for examples of the players caring in R&T play (Table 3.3). All of the selections were edited onto a separate tape from the master tape.

Table 3.3
Metacommunication & Expressions of Caring (Donaldson, 1976)

Hugging
Embracing
Touching
Smiling
Laughing
Talking
Eye contact

The participants were interviewed individually by the researcher twice and as a group once following the individual interviews. In each interview, the subject watched videotaped R&T play while responding to the researcher's questions. The researcher began the first interview by asking the participant to describe an R&T play scenario seen on the videotape. Further questioning was based on the direction from the boys' individual interpretations of isolated R&T episodes and Smear seen on the videotape. (It should be noted that the researcher did not have a set of questions that all participants responded to). In the second interview (following the same procedures as in the first interview), the participants discussed events that the researcher believed are indicative of caring for one another occurring in R&T play and Smear. The third and final interview involved all of the participants reviewing and discussing the tapes they observed separately without direction from the researcher.

FINDINGS

The primary objective of this study was to determine specifically the way in which caring was expressed in R&T play in a natural setting. The results of this study clearly indicate that R&T play is a way in which boys care for one another and this expression of caring is evident in the data. In total, 59 episodes of R&T play (Table 3.4) were observed and recorded. Another 60 R&T type behaviors were exhibited during Smear (Table 3.5). There were also 74 instances of care expressed (Table 3.6) during R&T play. Approximately 69% of instances of care during R&T play were expressed during Smear.

Table 3.4
Examples of Rough and Tumble Play (Excluding Smear)

Types of R & T Play	Frequency
Light touch on various body parts	10
Shadow boxing	5
Chase & flee	5
Picking up and dropping a player	4
Throw dirt, sand, ice cubes	4
Kick the ball at players against a wall	3
Light kick behind knee	3
Light pinch	2
Gang tackle by 2 or more	2
Lightly kicking sand	1
Running while twirling streamers	1
Climbing up wall with feet after running start	1
Running up the sliding board and jumping off	1
Arm wrestling	1
Gentle push off sliding board	1
Playfully grab for feet on monkey bar	1
Pulling down a sliding board	1
Total	59

As a result of the interviews, several findings emerged that provided some answers to the previously mentioned research questions. The data shows that the seemingly helter-skelter play of Smear had some very definite purposes with guiding forces (or rules) that remained constant. The summarized findings from this study will be discussed in the following sections:

Rough and tumble play is stylized.
Rough and tumble play appears to be an excuse for physical contact.
Friendship plays a role in how boys care for one another.
All of the players knew or observed the rules.
The rules separated players from intruders.
R&T play and Smear rarely resulted in injury.

Table 3.5
Smear Actions

Types of Actions Exhibited During Smear

Tackling		26
Gang Tackle	12	
By Neck	6	
Tackled by one person	7	
Pulled on clothing	1	
Avoided Tackling		5
Threw ball to avoid	3	
Fell down to avoid	2	
Gang tackled when could have been avoided		4
Other actions exhibited during Smear		25
Chase and flee	13	
Push or shove	4	
Hide	3	
Juke	3	
Trip	2	
Total actions on behavior including avoidance		60

Rough and Tumble Play is Stylized

The first finding is that R&T play is stylized play that is symbolic of aggression but not real aggression. This may confuse the uninformed observer. There was a definite rhythm or pattern to this play that is reminiscent of a dance. Those who did not know the steps to the dance (or in this case R&T) play had difficulty playing with those who did. The play of Smear is stylized or choreographed to the point that certain actions indicate to the participants that a game is being played and it is safe. Many of these actions also helped to keep the person in possession of the ball injury free. Recognizing a "play face" was something all of the players were able to do. The play face signals through facial features such as smiling, laughing, and eye contact (Blurton Jones, 1976) that inform the players that even though they are about to get hit, they are still playing. Those who are non-R&T players cannot "read" the play face missing the nonverbal cues altogether. For example, an aggressive child ignores the smile of an R&T player and crashes into the unsuspecting player causing feelings to be hurt and possibly injuring the child.

There was often an invitation to R&T play that included faking, hiding, and playful taunting. Often a player would stand as though inattentive as a way to get someone to chase him. A player would hide behind trees and peek out as if trying to tempt someone into chasing him. There would be playful teasing as simple as a declaration that, "You can't catch me" that would often be enough to get the others to chase him. Accepting the invitation to play, coupled with recognizing the play face, allowed each R&T player to know that it was safe not only to play but also to make himself vulnerable to R&T play. During the interviews with the researcher, the points made above are underscored. Kevin states, "Well, what makes it fun is that we're all trying to work together to get somebody down. We're playing together." Brian further states, "One person jumps on him and throws him down and everybody else goes after him. So we're working like a team." Clearly working together and knowing what the others were going to do, as in a dance, is critical to the success of this style of play.

Rough and Tumble Play: An Excuse for Physical Contact

Among the participants of Smear there is a healthy respect for personal boundaries that is often expressed through physical contact that at times is quite intimate. The participants knew what was considered appropriate touch during Smear and R&T play but were less certain of what was appropriate outside of the play arena. This lack of certainty was shown in responses to the researcher's questioning regarding where it was appropriate to touch when outside the realm of R&T play (Table 3.6).

While observing Smear it became evident that physical contact was often more than just an incidental part of the experience. The larger, stronger boys would allow many others to grab their arms or legs and then drag them for several feet before feigning a collapse that would result in what Kevin described as a "puppy pile." After being tackled, the boys would often linger on the ground laughing with their bodies touching one another.

Other times boys would walk around with their arms around each other in between instances of chasing after the ball or each other. When asked about their behavior, the boys were likely to say that they were "hanging on" to each other. When pressed, they would agree that hanging on to each other was similar to hugging, but the boys would not choose that terminology themselves. They were certain however that the boys were friends and it was perfectly fine walking arm in arm in the context of play. While observing two boys walking with their arms around each other Perry stated, "They have their arms around each other because they are friends." Brian contributed, "They look like buddies because they are hanging around together." When asked if this type of touch is permissible out of the context of play the answers varied by age (Table 3.6). Linden thought it was just fine to put his arm around anyone at anytime he pleased. Jake was noncommittal, claiming that he didn't know. Kevin was positive that any place out of the Smear or R&T play arena was not a good place to have his arm around another boy.

Table 3.6
The Setting in Which it is Permissible to Touch

Setting	Age in Years of Each Boy Responding				
	6	7	7.5	8	9
School	Y	Y	N	N	N
Store	Y	N	Y	N	N
R & T Play	Y	Y	Y	Y	Y
Movie	N	N	N	N	N
Walking	Y	Y	N	N	N

Y = yes   N = no

Rough and Tumble Play Goes Beyond Physical Contact

Boys caring for boys was evident throughout this study. For the purpose of this paper, intimate touch that would not normally occur, except in R&T play, is considered to be an expression of caring. When asked to review the video sequence where boys were walking around with their arms around each other, hugging, and touching hands they had this to say: Linden described this as "Being friendly." When Brian observed this he said that, "They're pretty much like friends" and went on to say, "They look like they're buddies." Jake let it be known that there is no way he would permit someone to hug him if they were not friends to begin with.

Perry viewed a video sequence in which he observed his brother, Patrick, and Zach walking around with their arms around each other and surmised that this was occurring because they were "friends" and they appeared to be "happy." Jake also observed Zach and Patrick "hanging on" in the same manner and that this was a "good" thing to do among friends. While watching Jake and Zach walking around hanging on to each other, Patrick also determined that they were friends and this was an okay thing to do if you are friends and care for one another. Kevin also agreed that it's okay to put your arm around someone when they are playing but nowhere else.

The boys had a clearer sense of what was appropriate touch of the body by a friend. When shown a videotape of a boy being picked up by a friend who had his hand placed on his derriere, the informants all agreed that this type of touch was okay as long as they were friends and just playing. However, it was quite another matter when asked what they would do if they were touched like that by someone who was not a friend. The older boys would be more likely to take matters into their own hands, whereas the younger boys would be more likely to seek help. Perry and Zach both claimed that inappropriate touch would lead to a fight. Brian and Kevin were far more poignant stating that "I'll kill him." Patrick simply said, "I won't let him do that." Linden said, "I would tell the teacher." Jake was also clear in what to do by claiming he would "Get him off."

Connection between the R&T players seemed to be paramount in Smear. For example, tackling or attempting to pull the person down who has the ball has little to do with the game of Smear itself. Connection was Linden's primary reason for playing Smear, consider his comments: "I just like being tackled." "You get to wrestle them down," and "You feel good" during this time. Perry had similar feelings about Smear saying he plays, "Because I just feel like getting tackled a little bit."

The Influence of Friendship on Caring During Rough and Tumble Play

It was apparent to this researcher that the intensity of physical contact marked differently for those who were considered friends. With those identified as friends, the play scenario lasted longer. With their friends, they gang tackled and held onto each other long after the ball carrier was down. They only put their arms around those they identified as their friends. Clearly, the boys enjoyed touch from their friends that was only evident during R&T play.

Of the seven who played together, Kevin claimed they were all friends. Linden and Jake liked playing with their friends because they liked to be tackled. Linden also said "When we play Smear we get to be with our friends." Patrick expressed similar feelings and remarked, "We are all friends and we play together all of the time." Brian said, "Most are my friends, some aren't. It doesn't really matter. We are having fun." Linden prophesied that "All of us need to look out for each other. We need to see if someone is hurt because we are friends." Linden went on to say, "If you want to be friends, you've got to help your friends and stuff." Zach concurred admitting that when he plays Smear, he is able to be with his friends. Zach further stated that he liked to play with them, "Because they're nice and like they listened to me like when I had my sunburn."

There were several things the players did to help their friends when they were shaken up (Table 3.7). Most of the time, a delay of the game sufficed while the player gathered his composure. Patrick was known to playfully fake being hurt but each and every time he went down, someone was always there to check on him. When asked why they want to check on each other so much, Perry thought for a moment and said, "Well, we're best friends" and "we're like brothers." Zach added, "Well if they get hurt we take them to a grownup" and as to why he did that Zach simply stated, "He's my friend."

Table 3.7
Types and Instances of Care During R&T and Smear

Caring Experience	Frequency
Physical	
Helped the fallen player up	7
Hugged and/or had arms around shoulders	5
Lingered on the ground after a tackle	5
Brushed off clothing	4
Jumped up and down with hands on another's shoulder	2
Balanced player before giving a gentle shove	1
Placed hand on buttocks	1
Held hands	1
Subtotal Physical	26
Non Physical but Close	
Checked to see if he was okay	11
Stopped the game until a player caught his breath	9
Held game up while boy composed himself	8
Called time out when a player fell accidentally	5
Helped find lost items	4
Walked with a player while he regained composure	3
Picked up glasses	3
Called for or took the player to the teacher	2
Took hurt child to teacher	1
Called for teacher	1
Subtotal Non Physical	48
Total Physical and Non Physical	74

In summary, friendship was extremely important in establishing caring-intimate relationships among the informants. Smear and other types of R&T play became a vehicle through which friendships grew. Those who were not friends or those who were overly aggressive may have been able to play Smear but were denied intimate contact by the core group of players. Friendship also accounted for the act of forgiveness. Patrick was pushed rather aggressively by Kevin into Zach. When asked why he did not retaliate, Patrick simply replied, "Because we're friends." Kevin was very concerned because he knew that he pushed a little harder than usual and went to see if he was all right.

Rules are Specific and Adhered To

Although unwritten, Smear had some very definite rules that governed conduct and behavior. The unwritten rules to Smear included the following; (a) wait
for a player to get up before play resumes; (b) when a player falls down accidentally, the fallen player retains possession of the ball; (c) if someone gets hurt
he will be tended to by a player and if needed a teacher is called for; (d) wait
until the ball is thrown and caught before tackling a player; and (e) a player
taking unfair advantage of a fallen player will be chastised by one or more
members of the group. Since the rules for Smear are not written down, players
recommitted to them daily. There were times when a new rule governing the
game was created. However, the new rule rarely lasted for more than a day and
was forgotten the next.

The players' knowledge of rules of Smear centered around the protection of
the players' physical well-being. Safe zones were created to keep players from
being tackled along sidewalks, trees, or other obstructions near the playing field.
When a player needed a moment to gather himself, a simple declaration of
"time-out" was mutually respected. When a player was inadvertently knocked
down or seemed to be slow to get up, at least one and often several players
would gather around to see if he was okay or if an adult was needed.

Keeping Out Unwanted Players

Anyone who wanted to play Smear was initially accepted even if the player
had a reputation for being aggressive. Two boys with whom the core group did
not like to play Smear with were Marcus and Daniel (who also seemed unpopu-
lar with most of the other children at the Youth Center). The pair had a reputation for hitting too hard and not slowing the game down when someone was
hurt. On one occasion, when Marcus and Daniel attempted to enter into Smear,
the two boys were summarily ignored by the other players. The uninvited and
unwelcome twosome soon left for another play area.

The intensity of contact was governed by the quality of friendship among
the players. How the Smear players treated fallen comrades was also influenced
by how they felt about them. There were no situations in Smear in which the
rules were not followed. However, the value of following the rules were not lost
on the players. When asked why it is important to play fair, Kevin acknowledged that, "Friendships might be lost and feelings can get hurt."

Aggressive players were treated as though they had an unwanted disease.
Paradoxically, they were tackled harder than those who followed the rules.
When asked what he would do if a person not following the rules fell, Patrick
said he would, "Just let him lay there and get stepped on." Brian had similar
thoughts on the matter although he was a little more civil claiming, "We don't
get real mad," and "If somebody's not playing fair sometimes we tell them not to
do it." Smear players will also try to stay out of the aggressor's way and as a
group will first warn, then ask the aggressors to leave if their play gets too

rough. As a final way to get the aggressive player to leave, the Smear players would stop the game until the unwanted player left.

Aggressive players who got hurt could expect not to receive an expression of care or offers of assistance. The game would not stop for the unwanted aggressive player nor would they assist him in any other way. Other things they might do to aggressive players include: push them away, step on them, laugh when they fall, leave them alone when they got hurt, let them walk by themselves, and try to stay away from them.

The Smear players knew the difference between the "play face" and an "aggressive" or "hostile face." They evaluated eye contact, facial expressions, and body posture to help make this determination. Each one of the informants easily identified what a person looks like when he is angry. Common descriptors included drawn lips, scrunched eyes, furrowed eyebrows, and running too fast. Kevin was able to further identify aggression when a person, "Has his hands turned inward and arms outstretched like he's going to choke you," Perry added, "His eyes look bad." Brian said this about the aggressor's eyes, "They're probably stretched out wide." Linden could identify an aggressive player, "Because they start pushing and stuff when they get mean." Jake claimed that an aggressor's eyes would be "weird," and "If they are knocking you down and taking the ball," that this would not be fair. Kevin recognized the vulnerability of the players of Smear and stated emphatically, "You're not supposed to like punch them in the stomach or anything like bite them." Kevin also took it upon himself to protect those in vulnerable positions. During an episode of Smear, Kevin angrily pushed an overly aggressive player who taunted and threw dirt on a fallen participant.

Injury Free

There is a belief among the Smear players that they are not likely to get hurt during Smear or R&T play. On the odd occasion when a friend is hurt, there is a nurturing response that is often missing in non-R&T play. There were essentially no injuries during R&T play and Smear. However, Brian stepped into a low spot in the sand and twisted his ankle. This injury was a result of running and not directly involved in bodily contact that is normally associated with R&T play. It also occurred at a time when no one was chasing him. Three other players came by to ask Brian if he was okay while another player recognizing that Brian was in pain, scurried off to get a supervisor.

The players believe that they will not get hurt playing Smear but also acknowledge that teachers do not like them to play, fearing someone will get hurt. Brian affirmed this by stating that when it comes to playing Smear, "Teachers won't let us." Kevin was of like mind stating that "Teachers sometimes don't understand how we play and they are afraid that we will get hurt." The fear teachers have, clearly seen by the boys, has been institutionalized in Early Childhood literature, most notably in the area of developmentally appropriate practice (DAP).

DISCUSSION

This study examined the roles of preadolescent boys engaged in R&T play and the way in which they recognized and expressed care for one another through R&T play. The findings from this research established through the eyes and words of the participants, that there is more to R&T play than what is often seen on the surface. Rough and tumble play is a paradoxical activity. It gives the appearance of aggression and hostility but in reality is an expression of caring. The R&T play of boys may look like a rambunctious free-for-all, spontaneous and chaotic, but in actuality it is a staged event, more like a rule-governed slow motion dance than a disorganized-spontaneous occurrence. Often the complaint about boys participating in R&T play is that it is unruly and contrary to good behavior. Outside observers are unaware of the rules governing R&T play. They see chaos and aggression, when what is really happening is a dance of controlled behavior.

The participants in Smear who identified themselves as friends had more intimate contact, protected one another from aggressive individuals, created their own rules for safety, and intuitively knew the difference between the "play face" of one who is friendly and the "hostile face" of an aggressive player. The boys also discussed how they liked Smear because it gave them an opportunity to "be" with their friends. For the participants in this study to "be," meant that they could "be" with their friends in a natural way without worrying about societal expectations or artificially imposed regulations. For these seven boys that meant touching, hugging, caring, connecting, and otherwise experiencing each other on a very intimate level.

To the participants, R&T play is a staging area for friendships, for negotiation, problem solving, to fulfill their need to belong to a group, to have intimate caring relationships with friends, experience friendly competition, develop a sense of community somewhere between the warmth and closeness of family and isolation, and indifference of the adult masculine world. However, there is a lack of support of R&T play in sources defining Developmentally Appropriate Practices (DAP). In the NAEYC handbook on DAP, Bredecamp (1992) calls for control of children who get carried away chasing, wrestling, or unduly trying to scare one another. Ward (1996) suggests that, even though aggressive play is often the result of short instructional periods, it should be eliminated. The DAP guidelines should include recommendations for playground space and time allotted for R&T play for those who choose this method of expressing caring in play. Conner (1989) found that most children are able to differentiate between R&T and aggression, however, it is the teachers who seem to have difficulty in separating the two. Perhaps teachers are more concerned with R&T leading to aggression as opposed to R&T in general.

Wegener-Spohring (1989) suggests prohibiting R&T would make children hide their active participation in games and alienate them even more from authority figures. Furthermore, teachers should learn the dynamics of active play and then would be in a position to assist children in playing more balanced activities on the playground (1989). In the opinion of this researcher, R&T play should be encouraged on the playground and playgrounds should be designed to accommodate R&T play. Teachers and administrators should be given training on the difference between R&T play and aggressive behavior and the nature of caring among preadolescent boys.

The frequently expressed concern that R&T play is too rough and somebody might get hurt was clearly not supported by the data. Brian, the only child to be injured in three hours of R&T play, mildly sprained his ankle when he accidentally stepped in a low spot on the ground. Furthermore, Pellegrini (1989) found that aggressive play amounts to less than 5 percent of all play on the playground. In this study, out of 119 instances of play that qualified as R&T play or Smear, there was only one injury that could be more accurately described as an accident. This amounts to less than 2 percent of play which is often considered aggressive and not permitted on the playground. However, this is not supported by the data. In this study, there were 60 incidences of R&T behavior that occurred in Smear (Table 5) that centered around tackling, chasing, fleeing, or otherwise trying to engage others into R&T play. This is the type of behavior that is typically observed by the playground supervisors which they feel compelled to control. Yet, injury as a result of R&T play accounted for less than 2 percent of the Smear and R&T actions combined. What the supervisors fail to see are the 74 instances of caring (Table 6) that is occurring during R&T play which, according to this study, is a primary way in which boys express care for one another. The amount of caring experiences far outnumber the R&T actions exhibited during Smear.

The participants in the study seem to have used the game of Smear and R&T play as a camouflage for the expression of intimacy for one another. Touching one another and expressing care for one another was evident in every R&T play scenario observed. Boys were hugging, walking around arm in arm, lying in what Kevin calls a "puppy pile" together after being tackled, and making sure the players were okay if they had inadvertently fallen down. Sports seem to be the arena in which intimate touch is socially acceptable for men and boys. Yet, this type of play is often denied for fear of injury among the players.

The unwarranted concern about safety may reflect more than the previously discussed fear of injury that causes adults to suppress R&T play. It could be that adults are uncomfortable with the sight of young boys hugging, rolling on the ground locked in each other's arms. This suppression of R&T play may be further illustration of our culture's homophobia (as the name Smear the Queer implies). As the data suggests, boys, as they get older, become increasingly uncomfortable with intimacy with one another. However, it is okay for boys to be intimate if the intimacy is disguised as something masculine such as tackling or wrestling.

Boys fully understand that touching each other is only acceptable in the context of play. This attitude is a reflection of our society. As adults, we too are only comfortable with males touching one another during play. Imagine for a moment two seven year old girls who are best of friends kissing each other as they part company. Now imagine two seven year old boys who are also best friends doing the same thing. The boys will likely be shamed for this behavior and taught that the masculine thing to do is "shake hands" and that boys do not kiss each other. This unwarranted fear of intimacy among boys may actually retard the emotional development of boys. Rough and tumble play appears to be an antidote to this homophobic induced suppression of feelings that boys begin to experience at an early age. The continued denial of R&T play may inadvertently contribute to the emotional unavailability of males and their inability to express themselves.

Opportunities for boys to express care for one another are limited. The older informants were keenly aware that touching was only acceptable at certain times and places. There were other times they knew touching was not permissible at all. This is what Gilligan (1997) is referring to when she said that their voices go underground. The voice Gilligan is referring to the ability of an individual to freely express their feelings and thoughts. Essentially, the younger the person is, the easier it is to express their feelings. The data in Table 3.6 demonstrates how the older boys were less enthusiastic about public displays of affection, whereas the younger boys were more likely to think it is okay to experience intimate touch in public areas. Yet caring, compassion, and connection are important for all people to understand and value (Noddings, 1993).

This study supports the idea that a boy's style of caring is revealed through a more socially acceptable manner such as R&T play. When boys are denied the opportunity to experience R&T play, they are also being denied one of the few socially acceptable ways in which they can express care and experience intimacy for one another. Alice Miller (1983) refers to this as "poisonous pedagogy." In other words, teachers are doing harm to children supposedly in their best interest. There was no other time when boys had or took the opportunity to interact with each other on an intimate level as they did during R&T play or Smear. It could be that we are retarding boy's emotional growth by denying them one of the few ways they perceive as appropriate to be in physical contact with one another. The most intimate form of play in this study was called Smear the Queer, seemingly named so in order to give themselves permission to be in close physical contact with one another thereby reducing the risk of being ridiculed. The data also suggests that further research is needed to determine the connection between how intimate touch is perceived with the age of the child being the determining factor. The question of whether or not aggressive children can be trained to participate in R&T play is also an area that should be researched. Research of day care and school personnel's attitudes and values toward R&T play need to be assessed. Finally, R&T play among girls should be evaluated to determine what this type of play means to them.

This study and its findings are added to a growing body of literature which supports R&T play as beneficial for the participants. More specifically, this study describes R&T play as an act of caring between friends, adding this dimension to the previous view of R&T play as being cognitively and physically beneficial for the participants. Rough and tumble play, as well as other aspects of childhood, need to be examined in the context of childhood, rather than the context of adulthood. We must learn to evaluate the experience based on its meaning for the participants rather than the observer. Even as we move into the 21st Century we have not learned to respect childhood and the rights of children to be childish and play as children do.

Robin Moore (1992) sums it up succinctly by stating why it is important to provide the space and opportunity for boys and girls to participate in R&T play (1992, p.13):

> To achieve a competitive role in a global economy it will be necessary to have creativity, imagination, vision, and adaptability to ever changing environments with the flexibility to adjust (with good humor) to new positions at a moments notice.

All of those qualities exist in R&T play.

REFERENCES

Blurton Jones, N. (1976). Rough-and-tumble play among nursery school children. In J. S. Bruner, A. Jolly, & K. Sylva (Eds.), *Play: It's role in development and evolution* (pp. 352-362). New York: Basic Books.

Bredekamp, S. & Copple, C. (Eds.). (1997). *Developmentally appropriate practice in early childhood programs.* Washington, DC: NAEYC.

Bredekamp, S. (Ed.). (1992). *Developmentally appropriate practice in early childhood programs serving children from birth through age 8.* (9th ed.). Washington, DC: NAEYC.

Cole, M., John-Steiner, V., Scribner, S. & Souberman, E. (Eds.). (1978). *Mind in society: The development of higher psychological processes.* Cambridge: Harvard University Press.

Conner, K. (1989). Aggression: Is it in the eye of the beholder? *Play and Culture, 2,* 213-217.

Csikszentmihalyi, M. (1975). *Beyond boredom and anxiety: The experience of play in work and games.* San Francisco: Jossey-Bass.

Donaldson, F. (1976). Metacommunication in rough and tumble play. *Reading Improvement, 13,* 235-239.

Fagot, B. I. (1984). Teacher and peer reactions to boys' and girls' play styles. *Sex Roles, 11,* 691-702.

Gilligan, C. (1997). Remembering Iphigenia: Voice, resonance, and a talking cure. In E. R. Shapiro (Ed.), *The inner world in the outer world: Psychoanalytic perspectives.* New Haven, CT: Yale University Press.

Harlow, H. (1962). The heterosexual affective system in monkeys. *American Psychologist, 17,* 1-9.

Ladd, G. (1983). Social networks of popular, average, and rejected children in school settings. *Merrill-Palmer Quarterly, 29*, 283-307.

Lincoln, Y. & Guba, E. (1985). *Naturalistic inquiry*. London: Sage Publications.

MacDonald, K. (1992). A time and a place for everything: A discrete systems perspective on the role of children's rough and tumble play in educational settings. *Early Education and Development, 3*(4), 334-351.

Miller, A. (1994). *The drama of the gifted child: The search for the true self.* New York: Basic Books.

Moore, R. (1992). The child's right to play: Expression of article 31. *Proceedings of the American Affiliate of the International Association for the Child's Right to Play* (pp. 11-15). Arkansas: Southern Early Childhood Association.

Noddings, N. (1990). Ethics from the standpoint of women. In D. L. Rhode (Ed.), *Theoretical perspectives on sexual differences* (pp. 382-392). New Haven: Yale University Press.

Noddings, N. (1993). Caring: A feminist perspective. In K. A. Strike & P. L. Ternasky (Eds.), *Ethics for professionals in education: Perspectives for preparation and practice* (pp. 43-53). New York: Teachers College Press.

Pellegrini, A. D. (1988). Elementary school children's rough and tumble play and social competence. *Developmental Psychology, 24*(6), 802-806.

Pellegrini, A.D. (1989). Elementary school children's rough-and-tumble play. *Early Childhood Research Quarterly 4*, 245-260.

Pellegrini, A. D. (1994). The rough play of adolescent boys of differing sociometric status. *International Journal of Behavioral Development, 17*(3), 525-540.

Pellegrini, A. D. (1995). A Longitudinal study of boy's rough-and-tumble play and dominance during early adolescence. *Journal of Applied Developmental Psychology, 16*, 77-93.

Pellegrini, A. D. & Perlmutter, J. (1988). Rough and tumble play on the elementary school playground. *Young Children, 43*, 14-17.

Piaget, J. & Inhelder, B. (1969). *The psychology of the child*. New York: Basic Books.

Sutton-Smith, B. (1992). Commentary: At play in the public arena. *Early Education and Development, 3*(4), 390-400.

Ward, C. D. (1996). Adult intervention: Appropriate strategies for enriching the quality of children's play. *Young Children, 51*(3), 20-26.

Wegener-Spohring, G. (1989). War toys and aggressive behavior. *Play & Culture, 2*(1), 35-47.

Part II

Cross Cultural Play Variations

A major relationship between evolutionary and cross cultural approaches to play is that the latter can be used as an attempt to discover what if anything is universal across all human cultures and therefore perhaps permits an evolutionary explanation. In 1959 Roberts, Arth and Bush published an article, "Games and Culture," which was the first quantified attempt to related cultural forms to play forms in modern social science. They used the Human Relations Area Files which were anthropological digests of data on multiple human cultures in about 50 of which there was available play data. What they found was that games of physical skill were practically universal, but that games of chance and games of strategy were more limited in cultural distribution. In general those cultures with only physical skill games were less complex than those with the other two kinds of games. Games of strategy in particular seemed reserved for societies of high political integration and social stratification. In later research by Roberts and Sutton-Smith (1962) it was found that there were also differential child training relationships with each kind of game. In general societies with only physical skill had a greater emphasis on self-reliance training (they were more often hunter- foragers, pastoralists or agriculturists) whereas societies with games of strategy were high on obedience training and self-restraint (as is required in chieftain and city cultures with complex social class arrangements and high political integration). Societies with games of chance were somewhat anomalous being found in both simple and complex societies, but seeming more often to be associated with benevolent gods and survival uncertainty as amongst nomadic peoples living in high latitudes. Most of the findings were replicated in a much later study using near 200 different societies (Chick, 1998). It needs to be pointed out, however, that these sources of data were all pre television, and there is some evidence that access to that source of sporting knowledge can make a considerable difference in play diversions.

There are, however, relatively modern parallels between the distinction of the kinds of play based largely on self reliance and physical initiative and the kinds based largely on obedience and imaginative initiative. In New Zealand of the nineteenth century, for example, most of the children were connected in some way with farming in the years prior to universal schooling (Sutton-Smith, 1959, 1981). Their play involved much hunting and fishing, exploring the wilderness, management of animals in addition such home made toys as catapults, bull-roarers, slings, bow and arrows, peashooters, popguns, water shooters, darts, kites, whips, sledges, fires, explosives. These accounts call our attention to children's self reliance and their capacity for innovation in their rural environment. They do give support to the notion that children used to be more enterprising in these ways in the nineteenth century. Although as has been seen, it was an enterprise with respect to natural things and the outdoors rather than an enterprise in matters of a symbolic order which was to come later. And again it was an enterprise which was associated with a harshness and aggressiveness of life, much of which was also to disappear. American accounts of the early pioneering period in similar self-reliant but somewhat rugged terms are those by Howard, 1977; McClary 1997; Mergen, 1982 and of course Mark Twain. In New Zealand when these rural children were sent to compulsory public schools in 1877 there was added a new and strong diet of playground games, the kind that has been written about in children's folklore (Sutton-Smith, Mechling, Johnson & McMahon 1999), such as Bar the Door, Botany Bay, Buck Buck, Bull in the Ring, Buttons, Chasing, Cat and Mouse, Cat's Cradle, Chain Tag, Chuckstones, Hopscotch, Jump Rope, Marbles, Caesar etc. (see the hundreds of further examples in Sutton-Smith, 1959).

While any simple typification of so many rule games is likely to be over simplified, one thing the children were getting was a thorough introduction into group rule obedience. You can't play these games if you don't follow the rules. And if you go through six years of such play and game regimentation you are probably more suited to an adult life, which is similarly regimented, as in the army, in all industrial jobs, in all commercial jobs etc. Once schools begin you are more enmeshed in collective obedience than you were as a solitary farmer's child. As the twentieth century advanced, however, the required kinds of obedience were less desired for these rule bound manual or physical types of work and more for proclaimed for the more literary and symbolic types as in non-manual deskwork. One may still play physical sports but their dominant place has now been challenged by all kinds of board games of strategy and chance and by the modern video and computer games which are mixtures of strategy, chance and narrative. Though there are occasions where the children themselves reconvert these video games back into forms of physicality. Thus Dorst (1999) observed this kind of physical game transformations amongst a group of second grade boys on a public playground in Laramie, Wyoming in 1990 in a game called 'Super Mario 5.' In this game, he says, the participants devoted much recess time to moving around the playground and acting out the sort of encounters and events that characterize the video game super Mario brothers, then the most popular electronic game among these boys.

> In their roving play they encountered and surmounted obstacles and barriers, fought a variety of dangerous creatures, acquired 'artifacts' that gave them enhanced powers, entered 'Warp zones' that allowed special movement and traversed boundaries from one imagined world to another. Though the ultimate object of the actual Nintendo game is to rescue a princess, the playground game was not particularly goal oriented. The point seems to have been the imaginative, improvisatory elaboration of the video games structure itself. And that means in their play the boys in-

cluded such things as putting themselves 'On Pause,' freezing the action as one can do with a button on the electronic control pad. This allowed for trips to the bathroom and other diversions. Also, they would devise theme melodies for the various 'worlds' they created, humming the appropriate tune as they moved about the imagined space of 'ice world,' for instance (Dorst, 1999, p. 270).

So much for adults who lament the end of children's physical play and imagination! But, as Dorst notes, "In many respects this game is not so remarkable. Children's lore offers many instances of the vernacular recontextualization of media form and texts" which Dorst calls 'double voicings.'

Human bodies after a million or so hominid years are still largely agents of self-reliance and for many moderns, the best forms of recreation continue to involve their bodies in somewhat similar exercises. What we have come to expect in children's play with decreasing physical self reliance and overt aggression, however is more obedience and self restraint, and with that the contrary increased stimulation for the management of symbolic information as well as a capacity for the exercise of variant private imagination which as we have just seen can itself convert imaginings into novel physical performances.

CHAPTER 4. CAROLYN POPE EDWARDS: CHILDREN'S PLAY IN CROSS-CULTURAL PERSPECTIVE: A NEW LOOK AT THE SIX CULTURES STUDY

What we have in the Edwards study is again an attempt to relate kinds of culture to kinds of play but this time she uses a small intense sample that represents different areas of the world and different complexities of social organization. Again, in the more complex societies with more children, more cultural events and more toys available and more free playground time, there is more time for play including more competitive rules games and more fantasy games. In Edward's more rural play cultures the children mostly have their time filled with chores, and in female cases this often results in playing role games (mothers, gatherers etc) which reflects their imminent employment. It need to be clear that there have been long periods in history when females had relatively little access to the dominant kinds of play life and there is probably a relationship between such play loss and lesser access to general political power roles. Social play is by contrast a constant imaginary access to and therefore rehearsal of power roles. In addition, however, there is generally less competitive play in these smaller survival groups, which depend on much greater direct mutual cooperation to survive. Still although there are differences in play complexity between the groups, Edwards also reports that everywhere children would occupy themselves at times with making things with mud and cloth, with self expression, with exploring, with imagining, with imitative rehearsals, with problem solving and with peer hierarchical collaborations. The picture we seem to get, as in the evolutionary materials of the prior section, is that almost everywhere there is a baseline of playfulness including at least play fighting, locomotion, social play and object play but beyond that each adult society also has differential cultural stance towards child play as well as towards what is permissible for boys and girls. For example: older forms of play are often perceived as neglectful by modern investigators although this may be mistaking an implicit induction of self reliance by the parents who are also making the children fend for themselves (Lancy, 2001), sometimes play by both sexes is strongly negated by parents because work or religious demands are exacting (Miracle, 1974), sometimes parents do play with the children in culturally constraining and socializing social or educational forms in order to advance the children in adult social and academic requirements. This seems more likely in complex ancient societies such as China and Korea (Goncu, 1999).

Sometimes in individualistic modern societies adults provide increasing opportunities for autonomous peer play and for solitary home play (Cross, 1997).

One of the most striking examples of the way in which local culture controls child play comes from the long list of ethnographic culture-play studies which have characterized the publications of The Association for the Anthropological Study of play since 1974. Andrew Miracle found that amongst the Aymara of South America (Bolivia, Peru, Chile) the heavy dependence on cooperative work (with sheep tending and crop weeding) for adults and children was associated with a negative attitude towards any play by the children in the vicinity of the home. They were told that they were not adults until they could put aside such play. As a result, children tended to play out of sight, and thus play became as it were an arena for these separate private play feelings. This out of sight play was sometimes encouraged surreptitiously by older women. The Aymara themselves were noted for their ability to maintain a totally non-emotional stance in the presence of their political overlords and to indicate agreement with the demands being made upon them without any intention of ever fulfilling such demands. It is not, therefore, too far fetched to think of their long years of private peer play secrecy as being disciplinary for that kind of adult mindedness where private feelings are completely reserved. In sum it is not improbable that the child play cultural socialization as a whole, was a framework for initiating the kind of relationship to work subjectivity that would be useful to them as adults.

Thus it is arguable, as we have said above, that the heavy diet of folk games (hopscotch, jump rope, bar the door, punt back) that once characterized the public playgrounds of the 19th and 20th century western world were in their intricate group and physical character a useful socialization into the highly intricate interactive manual labor which was typical in factory life. Again while moderns generally think of themselves as giving more freedom to children and more stimulation as compared with many other cultures, it is arguable also that all the endless playing with private toys is an example of socialization into the kind of solitary desk work and decision making that is required in the modern world. Which is to say that on the one hand, there is a basic kind of effectively instigated child play and on the other there is typically some adult socialization systemic within which such play is typically incorporated and transformed.

CHAPTER 5. JEAN PIERRE ROSSIE: CHILDREN'S PLAY AND TOYS IN CHANGING MOROCCAN COMMUNITIES

With Rossie's work, we have a chance to inquire more carefully into the role of toys in child development within culture. His chapter was originally presented as a part of a *Cultures of Toys* Symposium at Emory University in 1997 and is summarized elsewhere (Sutton-Smith 1998). The somewhat naïve issue before the conference was whether it might be advisable to bring to the USA for a Smithsonian outdoor folk exhibit, examples of the inventive toy play of children from preindustrial cultures in order to show modern children the greater creativity these other children had in making their own toys. Of the various speakers, Dr. Rossie had spent the longest span of years studying this phenomenon in such a society, which in his case was in parts of Morocco. The case he makes quite strongly is that the children in his groups invent toys often as a part of the recurrent festivals in that community (water pistols, windmills, swings, tambourines) but they also make toys from natural materials such as reeds, stones, squashes, potatoes, etc. which can be made up as dolls or as doll houses. The toy making in both cases is embedded in the larger communal festival or cooperative peer events, and once the events are finished, the toys are disregarded. He contrasted those practices with the emergent availability of modern consumer toys with which these children are also learning to play but which serve their private imagination rather than the social imagination as in the former case. He shows that these consumer toys do survive beyond the single playful occasion. These children are clearly caught between two worlds, that of the collective festival cultures and that of the individualist consumer cultures.

The contrast between the social imagination in the one case and the private imagination in the other seems useful. There is already data to suggest that modern children tend not to maintain an interest in their consumer toys unless they can incorporate them in their own fantasies (Sutton-Smith, 1992), and this suggests in turn that although they do not create these toys like the rural Moroccans, they do invent their own fantasies about them. Their creativity goes into their multiple fictions about the toy, whereas, the Moroccans perhaps invent the toy to subserve the preexisting fantasy of the festivals. The latter's toys do not outlive the festival but the modern toys sometimes remain for life either as antiques or to continue fuelling the reminiscences of childhood. For many modern adults, their toys remain as their companions throughout the rest of their lives. It makes sense to see that these kinds of toy fantasies are just one aspect of the intensive role of the variant private imaginations in modern life as supported by films, television, malls, books, games and toys. Private daydreaming and vicarious spectator play participation are incredibly diversified in modern life as compared with the legendary collective fantasies of the preindustrial world. The modern culture-play systemic is once again that of the desk person at home or at work dealing with great varieties of symbolic material as the normal requirement of the work place. The child's play world with private toys and multiple fantasy worlds for themselves is a reflection of the same kind of job possibilities in the non-manual world of the adults.

CHAPTER 6. KALYANI DEVAKI MENON: AT WORK AT PLAY: TRAINING HINDU NATIONALIST WOMEN

There is an almost frightening adult-child play systemic in the situations described in this chapter. Here as in so many other places the physical games periods are used supposedly to reinforce the larger prevailing unidimensional cultural beliefs. We need add that the kinds of games discussed by Menon are very general throughout the world of youth groups and are hardly distinctive of any single nationalistic philosophy. What is particularly interesting is to see that some of these very physical and very aggressive games are not unlike the play fighting described in Reed's rough and tumble study in the earlier section. The key actions can involve: knocking another player off her balance while hopping on one foot; groups of five in a circle bashing into other groups to knock them apart or off their feet. Most of the other games involved either a quickness of reaction or speed in catching the quarry. Yet the players felt they were more together at these times than during the non-playful training and thus one can again justify the Fernandez (1986) view that a major function of playing is bonding. As in the rough and tumble case for boys, it is likely that all this action not only released aggression but also released affection which was the dualistic character of the rough and tumble described earlier. But the larger literature on play fighting shows that sometimes aggressive children turn it into violence, and sometimes it is used as an excuse for bullying ("We were only playing") and, as in the present case, it sometimes allows the authority herself to be attacked so there is a complexity to this mess with different individuals getting different satisfactions from it as in most cases of group play. What should not be underestimated, however, is the fact that these key play performances involve attacking the other players more or less forcibly, and while little is said about such violence when the main patriotic doctrines are being presented during the rest of the day, there is very strong implicit agonistic sentiments implied by the doctrinaire rejection of the other supposedly "inferior" religions. In this sense, the aggressive games represent the implicit agonistic belief system better than the relatively controlled patriotic assertions and behavior of the major day. But what the play does is only "represent" those agonistic feelings rather than fully express them, which is the general nature of the play. It permits the officials the defense that this is not really aggression, when in fact it both is and is not aggression, which is the general paradoxical character of the relationship of play to all emotions and to all the otherwise supposed clarity of the non play belief system.

The fact that these "rough" games are being played here by females perhaps needs a further note because this kind of behavior in sports has usually been attributed to males only who are said by some to see their bodies as weapons or as instruments of power and violence for the dominance of females (Messner,

1990). Missing from the discussions of such behaviors is the fact that even as far back as medieval Europe, women participated in games that often generated into fights: "peasant women pushed, shoved, kicked and frolicked with as much reckless abandon as their fathers, brothers and husbands and sons; and they seem to have suffered as many broken bones and cracked crowns as the men did" (Guttman: 1978, p. 126). Guttman also notes that sometimes men were excluded from the rough games that only women were allowed to play.

CHAPTER 7. DONALD E. LYTLE: LUDIC PATHOLOGIES AND THEIR LINKS TO PLAY: CULTURAL AND NEUROCOGNITIVE PERSPECTIVES

The prior chapters have been illustrating the intricate ways in which play and biology and play and culture are intermeshed. Lytle's study of the Latah is even more striking. Here an event, the startle reaction, which continues uncontrollably in some persons (female, but not always) and which moderns might regard as a purely neurological symptom, is, in fact wedded from its moment of origin into cultural consequences of myriad kinds. The carriers may blurt out obscenities, may mimic the behavior of others, lose consciousness, sweat profusely, and, like the traditional cultural tricksters throughout the world and history, upset the usual expectations of order, decency, courtesy and poise. As Lytle says, teasing, joking and shocking humor and insults are the hallmark of the play forms. Latah is known by different names throughout the world; for example, in the Sudan and in Egypt among Arabic-speaking people it is known as *zar* where the trances and behaviors include dances with colorful scarves. Further, in Egypt among the youth, *zar* became a form of entertainment for young adults who 'put *zar* on stage' so to speak. By staging *zar* dances in discos some of the dances were actually based on the movements of *zar*-possessed women in traditional communities. Not unremarkably the same emotion of shock which is apparently the key emotional root for these symptoms, often gives us here the same medley of extremely denigratory playful behaviors that are found in analyses of initiation rites and hazing where those who are subjected to the occasion are treated in a multiplicity of dreadful ways just as are here exercised on others by the Latah subjects themselves (Houseman, 2002; Nuwer, 1999). Whereas in hazing the denigratory play forms (they always claim they are only playing) are meant as an initiation into the superior status groups involved, in the Latah case, the subject is induced into using similar outrageous "hazing" excesses for the entertainment of others and in a sense gains certain unaccustomed status through the centrality of the nonsense that is performed. Much milder forms of the same kind of play in which the central game position requires the player to suffer somewhat occur in Western folklore in such games as Blind Man's Buff, Dodgeball, Poisonball, King on the Hill, and Sardines (Sutton-Smith, Mechling, Johnson & McMahon, 1999). It is important to realize once again that these kinds of denigratory playful uses of extremity are a part of world cultural history and a permanent check on the notion that only the more idealized forms of play found in closely monitored modern nursery schools are what play is really all about. This idealization of the latter is simply not the universal nature of play as these Latah and hazing examples and our later discussion of the legendary play of "tricksters" will illustrate. That is, the list of play emotions above in the introduction to part I already makes it clear that there is a great deal of play that is dark, non rational, and even cruel and that these can be as much a part of the exciting nature of play as the many other instances of more lighthearted fun.

Chapter 4

Children's Play in Cross-Cultural Perspective: A New Look at the Six Culture Study

Carolyn Pope Edwards

Revisiting images, data, and notes from classic ethnographic and observational projects is worthwhile for several reasons. The first is that these are primary sources with historical and documentary value, providing a window into past times and the lives of individuals and families in cultural context. A second is that early documents, images, data, and findings are collections of systematic information that may continue to yield significant new scientific insights when reexamined in the light of new theories and/or reanalyzed using new variables or techniques. Yet a third reason is simply to increase appreciation of the development of the scientific discipline and the work of its major contributors.

All of these rationales figured into my desire to reexamine the observational notes and data gathered in the 1950s for the Six Cultures Study (Whiting, 1963; Whiting & Edwards, 1973; Whiting & Whiting, 1975) and its sequel, the Children of Different Worlds project (Whiting & Edwards, 1988). The Different Worlds project involved not only the Six Cultures data but also the extensive observational data subsequently collected by Beatrice Whiting in Ngeca, Kenya. As well, it involved a huge body of spot observation and running record data collected by several of the Whiting's students and colleagues in the 1960s and 1970s and collaboratively analyzed for the 1988 publication presented by B. B. Whiting & C. P. Edwards, in collaboration with C. R. Ember, G. M. Erchak, S. Harkness, R. L. Munroe, R. H. Munroe, S. Nerlove, S. Seymour, C. M. Super, T. S. Weisner, and M. Wenger. To the Different Worlds project, Ruth and Lee Munroe generously contributed spot observation data that formed the basis of several other publications with their students and colleagues (e.g., Bolton, et al., 1976; Nerlove, Munroe, & Munroe, 1971; Munroe, Munroe, Michelson, Koel, Bolton, & Bolton, 1983). The Six Cultures and Different Worlds data represent the first systematic cross-cultural data sets collected in multiple cultures using standardized methods. Revisiting the Six Cultures and Different Worlds data sets represents an appropriate way to recognize and further acknowledge the intellectual debt owed to those who contributed most to the running record and spot observational methodologies employed, John and Beatrice Whiting and Ruth and Lee Munroe, as well as to the others who contributed to these data sets and to the stream of comparative child research still very active today.

These data have documentary historical value because they offer a substantial archive of notes and coded data focused on child and family life observed for twenty years between the mid-1950s and mid-1970s, in communities that have undergone immense economic, political, and cultural change during the ensuing decades. In the everyday lives of children, schooling is one obvious source of change. More children have access to more kinds and levels of education and training, and schooling and literacy have become more determining of their future outcomes and life success especially in the developing world. Furthermore, the educational experience for children has itself undergone dramatic transformations in recent decades with respect to instructional goals, methods, and conditions. The children and families in these communities are also less isolated today. They have greater exposure to more forms of media and mass communication, experience more novelty, and come into daily contact with more products and by-products of the industrial world. Thus, revisiting the Six Cultures and Different Worlds data allows a look at the daily lives of children in home and school contexts during time periods very different from the present day. In retrospect, we may understand better what was seen then and also understand better what we see in the contemporary period.

The archives also warrant further study looking at new variables, asking new questions, or using new analytic techniques. Theoretical understandings of child development have changed in recent years. We have more complex theories of social participation and see children as co-constructors and negotiators in their worlds. We have deeper insight into the interaction of biology, culture, history, and psychology in providing contexts for social development. We have more complex theories of gender development involving differentiated labels and concepts, self-identity and self-perception, preferences, and behavior. All of the domains of children's behavior have received intensive study, especially their play, language, and symbolic behavior.

CHILDREN'S PLAY WITH TOYS

An international conference on play with toys provided the particular occasion for the study to be described. The conference was intended to showcase children's creativity in toy-making worldwide and provide an interdisciplinary forum for scholarly exploration of the psychological-developmental, historical, economic, cultural, and technological contexts of children' s toy-making and toy play. As a student and close colleague of Beatrice and John Whiting, I was asked to revisit the Six Culture and Different Worlds data. Child play was an area of child behavior that the Whitings had explored but then did not intensively address in their published writings. Yet Brian Sutton-Smith, conference chair, had always spoken of the promise of the Six Cultures data to shed light on the relationship of normative play patterns to cultural and gender socialization (1974, 1975, 1997, 1998). He had hoped these data would demonstrate how play and games serve two seemingly contradictory goals: the cultural replication of dominant core values, and its reverse, the acceptable outlet to express and sublimate motivations in conflict with dominant values (Sutton-Smith, 1979; Sutton-Smith & Roberts, 1971).

At the Culture of Toys Conference, Sutton-Smith (1998) took a different tack, a broader and more evolutionary view, and concluded that play has multiple motivations and functions for child and human development. For example, repetitive physical play in infancy improves children's control of motor skills, while exercise play in childhood increases strength and endurance, and rough and tumble play serves to develop (especially male) fighting skills and controlled dominance (Pellegrini & Smith, 1998).

Play with toys has a particular role in this picture and is vital for enhancing and fostering symbolic knowledge "in the individual mind . . . to stimulate the child mind to further growth and development" (Sutton-Smith, 1998, p. 7). Culture and tradition are critical to how much and in what ways children are allowed and encouraged to undertake such imaginative and cognitively enhancing play. Play with toys is hypothesized to be "mediated through social interactions and social traditions" (Sutton-Smith, 1998, p. 8), in several ways. First, cultural norms determine whether the play will be stimulated or whether it will be neglected (depending on whether adults consider it a good thing or a waste of children' s time). Second, norms determine whether parent intervention will serve to conservatively preserve tradition, or instead to instigate and foster independence and autonomy in girls and boys. Third, economic and historical conditions are critical resources for both physical and intellectual stimulation for play. Children play more in contexts where they have models for what they can do, and hence they play more elaborately in complex, densely settled communities with schools and mass media. They also make toys more in contexts where the economy and material world provide them raw materials in the form of natural materials and best of all, "trash," that is waste paper, wire, bottle caps, buttons, scrap lumber, cloth, tires, glass, cans, and so on, that can be fashioned into things to play with. The more plentiful the materials, the more children' s imaginations and inventiveness seem to be stimulated. Toy-making is part of a dynamic process of culture change (Rossie, 1998), in both industrial and non-industrial societies. From community to community, and without any adult involvement, news of new play-things can spread from one child to the next, creating fashions, fads, and crazes in the local, regional, and now global cultures of childhood.

These three hypotheses about play with toys seem consistent with Sutton-Smith's early reading of the Six Cultures data and his interpretation that play and games in those communities could be predicted by social mediation and cultural tradition because play and games provided appropriate preparation for later childhood learning and adult life. Creative and imaginative play, he further hypothesized, were more relevant and evident in "open" versus "closed" societies —more valuable to foragers than to tillers, differently expressed in boys versus girls, and most common of all in those complex societies where children were exposed to more novelty and/or had more leeway to roam about distant from home and choose their playmates and companions (Sutton-Smith, 1974, 1975).

This paper was intended to examine Sutton-Smith's hypotheses by reviewing the Different Worlds findings and then revisiting the Six Cultures data. Beatrice Whiting provided her files on child play, memories about what the data seemed to show, and cautions about the limitations of the play data due to the observers' focus on social interaction rather than individual behavior.

SAMPLES AND METHODS: A BRIEF REVIEW

The cultural communities studied in the Six Cultures and Different Worlds projects varied greatly in terms of economy, adult work, settlement and residential patterns. In each case, field researchers attempted to select a "primary sampling unit" of families who shared cultural beliefs, values, and practices and who knew one another's children (Whiting & Edwards, 1988, p. 21).

The Six Culture data were collected on children aged 3-10 years in six samples, studied by field teams between 1954 and 1956 under the overall direction of John and Beatrice Whiting (see Table 4.1). These communities were located in Kenya, Mexico, Philippines, Okinawa, India, and the United States. All but the U.S. community were at the time of study still primarily subsistence agricultural with some contact with the cash or wage national economy, and farmers sold some of their produce to purchase goods, pay school fees, or pay taxes. In three of the communities, families functioned as part of economically complex societies with class structures and an occupational division of labor. Women's workloads were physically heaviest in the Kenyan communities, and lightest in the economically complex communities in North India and the United States. Settlement pattern also varied. In Nyansongo, Kenya, families lived on the largest, most isolated farms; in other communities, homes were clustered into hamlets, villages, or towns. Household size and availability of kinfolk varied from the polygamous Kenyan households with as many as eight separate dwelling units, to the small nuclear households of New England. Kinfolk in most samples lived on contiguous land. The New Englanders were most isolated from relatives and also had the fewest children per family (average 3, versus 7-10 in North India and Kenya).

Subsequent to the Six Culture Study, running-record observations were collected on children aged 0-10 years in six samples (Whiting & Edwards, Appendix A). Gerald Erchak in 1970-1971 collected data on 20 children aged 1-6 years from 15 farming households in Kien-taa, Kpelleland, Liberia. Sara Harkness and Charles Super in 1972-1975 observed 64 Kipsigis-speaking children aged 3-10 years in Kokwet, in the Western Province of Kenya. Thomas Weisner in 1970-1972 observed 24 urban Luluyia-speaking children aged 2-8 years living in Kariobangi, an urban housing project in Nairobi, and 24 rural counterparts living in Kisa location in the Western Province of Kenya. Beatrice Whiting in 1968-1970 and 1973 collected data on 104 Kikuyu-speaking children aged 2-10 years from 42 farming and wage-working or entrepreneurial households in Ngeca, a sublocation located in Central Province about 20 miles north of Nairobi, Kenya. Susan Seymour in 1965-1967 observed 103 children aged 0-10 years from 24 lower, middle, and upper caste households in or nearby the Hindu town of Bhubaneswar in the state of Orissa in North India. In all of these communities, a sample of children was

Table 4.1

Characteristics of the Six Cultures Samples[1]

Communities & Locations	Linguistic Affiliation	Researchers during Years 1954-1956	Type of Settlement	Sample (Children Aged 3-10 Years)
Nyansongo, Western Province, Kenya	Gusii, Bantu	Robert LeVine Barbara LeVine Lloyd	Farm homesteads, population 208	18 homesteads; 16 children, 75 minutes of observation per child
Juxtlahuaca, Oaxaca, Mexico	Mixtecan	A. K. Romney Romaine Romney	Barrio, Indian section of a town, population 600	22 households; 22 children, 79 minutes of observation per child
Tarong, Luzon, Phillipines	Iloco	WilliamNydegger Corrine Nydegger	Scattered hamlets, population 259	24 households; 24 children, 135 minutes of observation per child
Taira, Okinawa	Hokan and Japanese	Thomas Maretzki Hatsumi Maretzki	Village, population 700	24 households; 24 children, 74 minutes of observation per child
Khalapur, Uttar Pradesh, India	Hindu	Leigh Minturn	Town, clustered, population 5,000	24 households; 24 children, 95 minutes of observation per child
OrchardTown, New England, United States	English	John Fisher Ann Fisher	Part of a town population 5,000	24 households; 24 children 82 minutes of observation per child

selected for timed-observation. The Six Culture methodology was developed collaboratively by the senior investigators, the field teams, and the staff of the Laboratory of Human Development in the Harvard Graduate School of Education. It was published in the *Field Guide for a Study of Socialization,* by John Whiting, Irvin Child, and William Lambert, and the field teams, *Volume 1* of the Six Culture Series, in 1966. This methodology was modified by Beatrice Whiting for work with the colleagues who contributed to the Children of Different Worlds project (Whiting & Edwards, 1988). Behavior was recorded by trained members of the children's culture, who had thus grown up with the same system of verbal and nonverbal communication norms. In recording the focal child's social acts (event sampling), the observer followed the eyes of the focal child, identifying whenever possible not only the child's social interacts but also the event that invoked it and any response by a social partner.

The records were taken in consecutive English sentences, for later coding. Behavior coding involved judgment of the apparent intention, which often could be made only when the entire sequence of events was known. Before an observation was started, the date, time of day, exact location, people present, and activities in progress were recorded. Time records were maintained along the left-hand margin of the paper, with notes as to when people entered or left the interactional space. With the exception of Bhubaneswar, India, observations were limited to the daylight hours and were distributed over four or five periods of the day. In the Six Culture Study, each record was five-minutes in length; in the later samples, they were 15 minutes to one hour in length, depending on the community. Methods of training observers and achieving inter-observer reliability were roughly the same across communities.

Spot observation data were collected on children aged 5-7 years under the direction of by Ruth and Robert L. Munroe and their collaborators between 1967 and 1975. On designated days, and at set time periods during the day, the observer visited all of the sample homesteads in turn and scored one set of records per subject child: proximity to home; predominant activity; sex, age, relatedness, proximity and activity of all persons present in the child's interactional space; persons' social engagement with the subject; and whether the subject was being supervised by an authority figure. By this method of instantaneous sampling, a total of 140 children were observed between 8 and 20 times each in six sample communities located in Kenya, Guatemala, Peru, and United States. Ruth and Robert Munroe collected data in Vihiga in Western Province, Kenya. Sara Nerlove collected data in Nyansongo, Kenya, and two Spanish-speaking Ladino farming villages in Guatemala, called Conacoste and Santo Domingo. Charlene Bolton, Ralph Bolton, and Carol Michelson collected data in Santa Barbara, Peru, a Quechua-speaking town with a mixed horticultural and herding economy. Finally, Amy Kohl observed children in Claremont, California, a primarily upper-middle class English-speaking suburban community.

Previous Findings: A Review

An overarching conclusion of Whiting and Edwards (1988) was that it is far easier to describe gender, age, and cultural differences in children's typical companions and activities, than to find differences in their social behavior (relative proportions of nurturance, dominance, dependence, and sociability) after controlling for companions and activities. There were four strong findings that were consistent across cultures and upheld by both running record and spot observation data, and all are relevant to cultural and gender differences in children's play.

First, girls spent more of their day doing responsible or productive work, such as childcare, housework, and gardening, while boys spend relatively more of their time playing. These sex differences are seen from age three onwards. For example, in all six samples of the Six Culture Study, boys aged 4-5 years had a higher percentage of observations in which they were seen playing than did girls (see Table 4.2, column 2). Girls aged 4-5 scored higher than boys in five samples, except in Orchard Town where both boys and girls had zero observations scored at work; and girls aged 6-10 scored higher than boys in working in 4 of 6 samples (Whiting & Edwards, 1988, Table 2.11). The spot observation findings on children aged 5-7 showed boys higher in playing in five of six samples, and girls higher in working in four of six samples (Whiting & Edwards, 1988, Table 6.2). Children in the Kenyan communities, whose mothers had heavy responsibilities in subsistence agriculture, least help from husbands, and large family size, scored particularly high in working (animal care, child nursing, housework, and agricultural tasks).

Second, gender segregation was the grand rule of social interaction during middle childhood (age 6-10). Boys and girls segregated into same-sex peer groups whenever there were enough children available, and especially did so when they have already divided themselves into age-homogeneous groupings. After age six, schooling was a factor that allowed boys and girls in most communities access to peers. For children under six, settlement pattern made a big difference. For example, in the Six Culture data, an average of 66 percent of the interaction of 4-5 year olds with children was with same-gender children aged 3-10 (Whiting & Edwards, 1988, p. 231). Since the majority of children this age were not allowed to range far from home, they were limited to who was available in their choice of playmates. The three Six Culture communities with the highest percentages of same-gender interaction, were Taira and Juxtlahuaca where children played in public areas and the streets, and Nyansongo, where boys went off together to herd cattle (combining work with play) in the fields.

Third, during middle childhood, boys reduced contact and interaction with their mothers and other adult females, and were observed at greater distances from home than were girls. The highest levels of distant-from-home observations in the Six Culture Study were found in the 6-10 year-olds from the denser settlements of Orchard Town, Taira, and Khalapur, where school-aged children (the boys only in Khalapur) had freedom to wander to play (Whiting & Edwards, 1988, Chapter 2). The spot observation findings clearly corroborated that boys aged 5-7 had more autonomy and freedom to wander and play. Boys were observed more often than girls in "undirected activities" (e.g. play) in five of the six samples (all but Claremont), and at higher average distances in feet from home than were girls during "undirected activities" in those same five samples (Whiting & Edwards, 1988, Table 2.6).

Fourth, girls, especially during middle childhood, had more contact and interaction with, and responsibility for, infants than do boys. For example, in the spot observation findings, 5-7 year old girls were observed holding an infant more often than boys in all samples except Claremont, and engaged in childcare in all six. The Nyansongo children scored highest with 14 % of girls' spot observations scored as holding an infant and 21 % as doing child care (Whiting & Edwards, 1988, Table 2.12).

Revisiting Childhood Play in the Six Cultures Data

These findings suggest that girls and boys in different communities have different opportunities for play, but they do not provide much detail. More can be learned from the Six Cultures observations, because observations were coded for different types of play, and the data deserve a second look. Both the coded scores and original field notes were reviewed for the following analysis.

Four major play categories had been used in the Six Cultures codes to score children's observation protocols: *creative play, fantasy play, role play*, and *games with rules*. Each 5-minute protocol could be scored for one type of play (creative, fantasy, or role play); if more than one of these three types of play occurred in a protocol, the dominant type only was considered. However, it was possible for a protocol to contain both one type of play (e.g., creative) and also be coded as a game with rules. Creative play was defined as play involving manual dexterity, and will be referred to in this paper as "creative-constructive." Examples included drawing in sand, constructing an object like a slide or swing, whittling or carving, making a structure of sand or mud, knitting or spinning for pleasure, or playing with a store bought toy creatively. Fantasy play was defined as playing a fantastic person or animal or pretending an imagined character. Role play was defined as acting out an adult role that the parents might actually perform (such as smoking, cooking, cleaning, shopping, parenting a doll, dressing up, etc.). Bloch and Adler (1994) called this role play "play-work" in their excellent analysis of Senegalese children's activity; the young children displayed many instances of imitative "playing at work" that gradually and imperceptibly slid into serious work as they grow older. Games with rules included games of physical skill, chance, or strategy, such as hide and seek, marbles, tag, fencing, jump rope, checkers, card games, and so on.

Because Beatrice Whiting had less confidence in these codes than in the social interaction codes, a conservative and basic method of analysis seems most appropriate. Thus, a count has been performed of boys and girls in each community sample ever scored as engaging in each kind of play, without taking account of how many instances were scored (usually 1-3 per child). Such a procedure tends to flatten rather than inflate group differences, and so those that are found, it may be assumed, were probably there in the data, not a statistical artifact. The findings are presented in Table 4.2, Columns 2 to 5. These columns shows percentages of girls, boys, and all children in each sample combined, observed as engaged in creative-constructive play, fantasy play, role play, and games with rules.

The findings in Table 4.2 suggest that Nyansongo and Khalapur children scored lowest in play, Juxtlauaca and Tarong children in the middle, and Taira and Orchard Town children highest across the four categories. The findings will be

Table 4.2
Children's Play for the Six Cultures Samples

Percentages of Sample Children Observations

Community Location	Aged 5-7 coded as play	Aged 3-10 in creative-constructive play	Aged 3-10 in fantasy play	Aged 3-10 in role play	Aged 3-10 playing games with rules
Nyansongo, Kenya	Girls 17% Boys 20% B+	Girls 37.5% Boys 62.5% B+ Total 50%	Girls 0% Boys 25% B+ Total 12.5%	Girls 12.5% Boys 12.5% B=G Total 12.5%	Girls 12.5% Boys 12.5% B=G Total 12.5%
Juxtlahuaca, Mexico	Girls 42% Boys 59% B+	Girls 45.5% Boys 18.2% G+ Total 31.8%	Girls 0% Boys 0% B=G Total 0%	Girls 31.8% Boys 9.1% G+ Total 36.4%	Girls 18.2% Boys 9.1% G+ Total 13.6%
Tarong, Phillipines	Girls 43% Boys 55% B+	Girls 33% Boys 41.7% B+ Total 37.5%	Girls 25% Boys 75% B+ Total 50%	Girls 41.7% Boys 83.3% B+ Total 62.5%	Girls 58.3% Boys 66.7% B+ Total 62.5%
Taira, Okinawa	Girls 79% Boys 92% B+	Girls 58.3% Boys 58.3% B=G Total 58.3%	Girls 33% Boys 33% B=G Total 33%	Girls 83% Boys 25% G+ Total 54.2%	Girls 75% Boys 75% B=G Total 75%
Khalapur, India	Girls 18% Boys 34% B+	Girls 41.7% Boys 16.7% G+ Total 29.2%	Girls 16.7% Boys 25% B+ Total 20.8%	Girls 16.7% Boys 25% B+ Total 20.8%	Girls 8.3% Boys 33% B+ Total 20.8%
OrchardTown, United States	Girls 43% Boys 55% B+	Girls 58.3% Boys 66.7% B+ Total 62.5%	Girls 41.7% Boys 25% G+ Total 37.5%	Girls 33% Boys 8.3% G+ Total 41.7%	Girls 75% Boys 50% G+ Total 62.5%

Table 4.2 condensed from Whiting & Edwards (1988), Tables 2.1 and 2.2.

discussed culture by culture, to explore how play was encouraged or neglected in each of the communities, for both boys and girls.

Nyansongo, Kenya

This community evidences the least play among the Six Cultures samples. Children were absorbed into the work of their mothers, and they helped with agricultural work, animal care, and childcare, and were discouraged from leaving their homesteads in order to minimize aggression with neighbors. Neither their mothers nor fathers stimulated their play by joining in, making suggestions, or providing help or materials. However, the children played with their kin in mixed age groups, and often combined play into their work, which was usually not arduous. For example, much of their interactions with infants and toddlers involved imitation, laughter, and playful touching and teasing. In one observation, a girl aged 7 was home with the baby (no adult present) and she entertained the baby by pretending to make a cigarette out of a cornhusk. When boys were herding cattle and goats in the lineage pastures, likewise, they played in ways that could be interrupted if the animals wandered or needed attention. For example, they made "plows" and "hurdles" out of sticks. The mixed age play of Nyansongo children was not conducive to competitive games with rules, however, and only tag and dirt-throwing contests were seen (for an example of girls playing tag, see Whiting & Edwards, 1988, pp. 207-208). The children had only a few simple homemade toys, such as slingshots, and perhaps because they had real babies to play with and real adult work to do from early ages, they did little role play. Their highest play scores were in the creative-constructive area, especially for boys. For example, in observations on adjacent days, boys in groups of four aged 4- to 11-years-old, played in a stream and water hole where their cattle were watered, and built themselves a dam. The dam building required some cooperation and led to joyful swimming and shouting. One boy, age 6, was first hesitant to help because he thought there might be snakes in the water, but then reassured by an older boy, became so engrossed he would not stop digging even though the others shouted at him repeatedly that his cattle were getting away.

Khalapur, India

Khalapur was a community high in complexity and relatively low in play. The children, when at home, engaged in much idle sociability and standing about, since the courtyards were crowded and they were discouraged from rowdy play. Their mothers often interacted with them in a somewhat scolding or reprimanding style and did not seem to encourage their play. However, children had considerable leisure, as boys were not expected to take care of cattle before age 6, and younger girls did only light household work. Both girls and boys did watch the infants and toddlers while their mothers cooked.

Boys had much more freedom to roam, and the older boys, watched by the younger ones, played different games with rules in the pastures, such as jacks and forms of hockey. Children had few toys, but often played with sticks or bits of paper or cloth. In one observation, an older boy chopped branches off a stick to make a hockey stick; he had considerable difficultly cutting the stick but moved it from one stone to another, still working at the end of the 5-minute observation. His mother, sister, and a visiting girl were nearby, but no one interacted with him except his sister who scolded him (which he ignored). In another observation, a 6-year-old girl was trying to embroider a little piece of cloth. Her aunt and grand-mother were nearby but did not help her. She threaded the needle after many tries, and then followed the lines of a design, carefully counting the stitches and correct-ing her mistakes. After the 5 minutes, her aunt and grandmother inspected her work and told her to take it all out because she was ruining the cloth. The observer commented in her notes that "actually she had done quite well for a first try." The girl continued to work on and try to correct her mistakes.

Juxtlahuaca, Mexico

Juxtlahuaca was a community with intermediate play and work scores. Older girls were the group most involved in responsible work (such as child care). Younger children were kept inside their courtyards, where there was usually a supervising adult present. These adults did not stimulate or encourage play but they were tolerant and non-critical. Children ran errands for adults when asked and, in between, engaged in much unstructured play. Children in the courtyards were observed trying to play a few simple games with rules, such as tag and ball, and they loved playing with the dolls and other toys introduced by the field observers, the Romneys, who had their own children (Beatrice Whiting, 1998), personal communication). What the children were high in was role play and creative-constructive play, especially for girls. They made houses, pretended to sew and to make tortillas. They dug in the sand with various implements, braided and played with pieces of palm and cloth, pounded bottle caps, built "tents" out of shawls, played "lariat" with rope, and used old bricks in numerous ways. The boys held top-spinning contests and played with toy cars, and made roads and vehicles out of mud. Fantasy play was not seen, however.

Tarong, Philippines

Tarong was another community with intermediate play and work scores. The busy mothers used work as a way of keeping their children busy but also depended on organized group games. Adults were almost always nearby and could oversee their children without being obtrusive. Older children went off to school, but after school, were given younger siblings to supervise. The field observers, the Nydeg-gers, noticed how the older children took great care to teach the younger ones how to play a variety of games with rules, including hide and seek, tag, drop the handkerchief, and other games they learned in school, "junior versions of school games, fantastic games labeled basketball or baseball, but bearing little resemblance to the originals. As many as 20 children, ranging in age from toddlers to 10 years, were observed in such riotous games" (Nydegger & Nydegger, 1963, in B. Whit-ing, ed., p. 834).

Table 4.2 shows the children of Tarong, especially the boys, as being high in all four categories of play. Examples of role play included playing "smoking," planting and harvesting, cooking and eating, haircutting, ironing, pounding rice, and having sex. Examples of fantasy play included playing "ghost," "jeep," "train," "horse," "swordfight," "bicycle riding" (on a drum), "card playing" (with leaves), and performing music (marching, singing, playing "guitar"). For creative-constructive play, they were resourceful in using natural and found materials. For instance, they made mud pies, constructed miniature seesaws out of sticks, drew in the dirt, constructed houses of branches, made toy cars of cups and cans, made guns out of bamboo, and made whistles out of banana stalks. The observations are notable for displaying young children engaging in long, sociable, harmonious, and constructive episodes of cooperative play. In one, two younger boys went under the porch, where adults could not see them, to build a play-house. They sat about a foot apart, facing each other, using various branches, sticks, and other materials lying around to build with. They chopped sticks, then pounded them into the ground and built a platform. Humming, chatting, and singing together, they proceeded to carefully and laboriously balance sticks across their uprights, mutually making adjustments and building and rebuilding their platform, for the length of their observation period.

Orchard Town, New England

The children of Orchard Town scored lowest in responsible work and highest in access to toys and manufactured games and store-bought play materials. Both boys and girls had high play scores, with girls notably high in games with rules. Because they lived in small families with few adults and children, and did not have freedom to wander their neighborhoods, they spent many hours indoors. There, their parents (both mothers and fathers) encouraged them to entertain themselves by playing alone or together, and the adults sometimes joined in, answered questions and offered information, or mediated disputes as needed.

Table 4.2 shows that children evidenced all four categories of play. For role play, they played "birthday party," cooking, making beds, having baths, shopping with "money," and dressing and taking care of baby dolls. Their fantasy play was imaginative and included playing "sheriff and deputies," "riding horses," (using a ruler), telephone-calling, "magic carpet" (with a tiny rug), "radar" (a dog serving as the airplane, and a wire stuck into the wood as the radar receiver), and "people" (using marbles). For creative-constructive play, they played follow-the-dots, coloring, cut-outs, drawing, cutting Venetian blinds out of paper, building with blocks, cutting up magazines, making paper chairs, sculpting with clay, making sand pies, and playing with a doll house. They also had books and television to fill their time, and many board games and card games.

Taira, Okinawa

The children of Taira had the highest play scores of the Six Culture Study, and the girls had notably high scores, especially in comparison with girls from other communities besides Orchard Town. Their mothers and fathers were heavily involved in physical work, but the children had much freedom to wander and play in the open, friendly courtyards. Young children under age 5 were seldom used in chores, and they attended a community nursery school in the morning. There, the teacher taught them turn-taking and other skills conducive to playing games with rules. After school, the older children supervised the young ones; all playing together in large play groups.

Most of the children were seen engaging in all four categories of play. Like the children of Taron, they were resourceful in using natural materials for creative-constructive play: they drew figures and house plans in the sand, built slides out of pieces of tile, made mud pie trucks, wrote with chalk on the wall, dug gravel pits, and made houses of bamboo sticks. For fantasy play, they imitated a track meet, acted out a sword fight, played house (with a "robber") and telephone-calling, played vehicle games with all sorts of blocks and bits of wood, used gravel to make "rain," and played "ghost." Their role play included playing house, store, and animal care. For games, they played marbles and pitched rubber bands, wrestled and chased one another.

SUMMARY AND CONCLUSIONS

This paper has involved a qualitative and quantitative reanalysis of the data on children's play from the Six Culture Study. The data, collected in the 1950s using a running record procedure, offer a picture of worlds of childhood at a time when communities were more isolated and less involved in markets and modern technology than today. Children in many communities participated in subsistence activities and child care in ways that promoted their responsibility and nurturance but still allowed time for play with peers and siblings (often they combined moments and episodes of fun, entertainment, and constructive problem-solving into their work). Children also had less exposure to novelty and stimulation coming from media and recreational and educational institutions than today, and less access to the products, by-products, and waste products of the industrial world. Only in Orchard Town, New England, did children already in the 1950s have a plethora of games, art materials, and toys, as well as tools, scraps, and "trash" so useful to use and combine with natural materials in constructing toys and creating imaginative narratives.

As a result of these contextual factors, children's play showed much variability across the sample communities. Games with rules, for example, were more prevalent in the three complex communities where parents were part of classed societies with economic role specialization and hierarchy (Orchard Town, Khalapur, and Taira) and in Tarong, with its nucleated living arrangements, as Sutton-Smith and Roberts (1971) concluded. Since competitive games thrive best with peer groups, rather than mixed ages, they were most often seen on the school playgrounds rather than backyards.

Role play was quite common in most communities, and higher in girls than in boys in four samples (equal in a fifth). These findings agree with Sutton-Smith's (1974) hypothesis that role play allows children, especially girls, to prepare for customary adult roles that they are expected to assume, by imitating the easily-observed activities of the people around them. In many of the Six Culture communities in the 1950s, there was little emphasis on material possessions, and children were more interested in interacting with one another and participating in the life around them than in the physical culture. They used mud, sticks, stones, and other natural materials to imitate adult roles of cooking, grinding, and plowing, and tied rocks and rags to their backs as pretend babies. They also used knives, pangas, and axes to imitate adult work and food preparation, as early as age two and three. "Smoking" and "telephoning" seemed to be two adult pleasures that children avidly imitated. Small children also liked to play "school" in imitation of their older brothers and sisters. Privilege in the communities was seen as going with age, and adults controlled the environment through work activities, much of it visible to children.

It is interesting to note that role play was particularly low in Nyansongo, probably because children there participated earliest and most heavily in real adult work and therefore did not need to "practice" through acting out. The line between the categories of work and role play can be blurry for such children, anyhow. Children growing up in subsistence communities are observed from toddlerhood onward to engage in a kind of "playing at work" or "work-play" (imitation immediate or deferred) that allows children to ease themselves from a playful trying-out, to serious effort, to full responsibility, in a gradual, self-motivated way, as they master the required skills and contribute ever more to the household economy (Bloch & Adler, 1994). Indeed, in all the samples, what was coded as "role play" seemed to drop off at the age that children were required to play a significant role in the household. Among the older children, role play was most commonly seen in the children of Taira, where children were not expected to engage in work.

Fantasy play was most prevalent in Tarong and Orchard Town, followed by Taira, but was observed among less than half the children in the other three communities. These findings would seem to support Sutton-Smith's conjecture that imaginative play is more relevant and evident in "open" societies, where children are exposed to more novelty and stimulation and/or have more leeway to wander and play without close adult supervision or control and where they can choose their playmates and companions (Sutton-Smith, 1974, 1975). Imaginative play included the most "fantastic" scenarios (e.g. "magic carpet" and "radar") in Orchard Town. Today, I believe, fantasy play would be more frequent and elaborated in all of the communities. Beatrice Whiting (1998, personal communication) observed that the Juxtahuaca children loved the toys brought by the Romney children, and children in all the samples immediately liked the pipe cleaner dolls when used by the field teams for the Thematic Apperception Test (TAT).

Creative-constructive play, finally, was evident in all six of the communities. Children seemed to have a developmental need to make and combine things, to make marks and draw, and to handle and reshape materials that could not be subdued. Indeed, in observations described above from Nyansongo and Khalapur, children continued in their self-directed constructive activities with mud and cloth even when criticized or told to stop: they were simply too absorbed and interested to heed others' interventions. Although the Six Cultures observations did not offer examples of children making complex toys, such as dolls with costumes, bottle-cap figures, or wire cars, nevertheless the children were seen building houses, dams, and cars and roadways, and drawing in and on all kinds of surfaces. Probably, as they engaged in these creative-constructivestories, their minds were actively constructing "stories" or "event scenarios" to themselves, and thus either role play or fantasy play was taking place implicitly. In recent decades, as children have gained exposure to more models of other children's play, as well as more access to materials and resources for play, their variety and complexity of creative-constructive play can be predicted to have increased in the communities. One corroborating anecdote is the craze for wire cars which the boys of Ngeca, Kenya, were seen making in the 1970s (see Frances Cox' photo of a male child nurse, carrying his infant charge on his back while racing his wire car down the dirt road, on page 169 of Whiting & Edwards, 1988). Field observer, Carol Worthman, instituted the "wire car safari" to the Ngeca children, and boys of all ages from eight into the teens, were seen to take part (Whiting, 1998, personal communication).

Children in all communities, this review suggests, seemed to have an appetite for self-expression, peer collaboration, exploration, rehearsal, imagination, and problem-solving. Their outlets were socially mediated, and took many varieties of play and work, with not necessarily clear boundary lines dividing them in the communities where children were most helpful. Both play and work seemed to allow children to build their repertoire of skills and schemes and to exercise and extend their knowledge and control over their environments. Cultural norms and opportunities determined the degree to which play was stimulated by the physical and social environments. Key factors included whether adults considered play a good use of children's time or just an annoyance, whether adults preferred to conservatively preserve tradition or instead to instigate innovation, and whether the environment provided easy access to models and materials for creative and constructive play. Nevertheless, play of several kinds was observed in each community and depended on the environment only for reinforcement, not for instigation.

REFERENCES

Bloch, M. N. & Adler, S. M. (1994). African children's play and the emergence of the sexual division of labor. In J. L. Roopnarine, J. E. Johnson, & F. H. Hooper (Eds.), *Children's play in diverse cultures* (pp. 148-178). Albany, NY: State University of New York Press.

Bolton, C., Bolton, R., Gross, L., Koel, A., Michelson, C., Munroe, R.L. & Munroe, R. H. (1976). Pastoralism and personality: An Andean replication. *Ethos, 4*, 463-481.

Munroe, R. L. & Munroe, R. H., Michelson, C., Koel, A., Bolton, R. & Bolton, C. (1983). Time allocation in four societies. *Ethnology, 22*, 355-370.

Nerlove, S. B., Munroe, R. H. & Munroe, R. L. (1971). Effect of environmental experience on spatial ability: A replication. *Journal of Social Psychology, 84*, 3-10.

Pellegrini, A. D. & Smith, P. K. (1998). Physical activity play: The nature and function of a neglected aspect of play. *Child Development, 69*(3), 577-598.

Rossie, J. P. (1998). *Toys in changing North African and Saharan societies.* Paper presentedat the Culture of Toys Conference, Emory University, Atlanta, Georgia, January.

Sutton-Smith, B. (1974). Towards an anthropology of play. *Newsletter of the Association for the Anthropological Study of Play, 1*(2), Fall, 8-15.

Sutton-Smith, B. (1975). *The study of games: An anthropological approach.* Opening address to 550th Anniversary Conference of the Catholic University of Leuven, Belgium, April.

Sutton-Smith, B. (1979). The play of girls. In C.B. Kopp & M. Kirkpatrick (Eds.), *Becoming female: Perspectives on development* (pp. 229-257). New York: Plenum Publishing Corp.

Sutton-Smith, B. (1997). *The ambiguity of play.* Cambridge, MA: Harvard University Press.

Sutton-Smith, B. (1998). *How do children play with toys anyway?* Opening address to the Culture of Toys Conference, co-sponsored by the Smithsonian Institution and Emory University. Atlanta, Georgia, January.

Sutton-Smith, B., & Roberts, J. M. (1971). The cross-cultural and psychological study of games. *International Review of Sport Sociology, 6*, 79-87.

Whiting, B. B. (Ed.). (1963). *Six cultures: Studies of child rearing.* New York: Wiley.

Whiting, B. B. & Edwards, C. P. (1973). A cross-cultural analysis of sex differences in the behavior of children aged 3-11. *Journal of Social Psychology, 91*, 171-188.

Whiting, B. B. & Edwards, C. P. (1988). *Children of different worlds: The formation of social behavior.* Cambridge, MA: Harvard University Press.

Whiting, B. B. & Whiting, J. W. M. (1975). *Children of six cultures: A psycho-cultural analysis.* Cambridge, MA: Harvard University Press.

Whiting, J. W. M., Child, I. L., Lambert, W. W., Fischer, A. M., Fischer, J. L., Nydegger, C., Nydegger, W., Maretski, H., Maretski, T., Minturn, L., Romney, A. K. & Romney, R. (1966). *Field guide for a study of socialization.* New York: John Wiley.

Chapter 5

Children's Play and Toys in Changing Moroccan Communities

Jean-Pierre Rossie

INTRODUCTION

First, I want to stress that it is impossible to claim any representativeness and completeness of the gathered information on Moroccan children's play and toys. The information describes existing play activities and toys but cannot be used to prove the non-existence of other games and toys. The research fields and the involved families have mostly been found through the chance of fortunate contacts and I here want to express my sincere thanks for the hospitality and collaboration received from many families and individuals, especially primary and secondary school teachers.

Before analyzing Moroccan children's play activities and toys, the reader may find some usefulness in the following notes on my scientific work, sources of information and research methods. During my earlier studies my major topic of interest has been childhood and socialization. Between 1975 and 1977, I did fieldwork among the semi nomadic Ghrib from the Tunisian Sahara. It is during my first field trip that I experienced the advantage and usefulness of participating in children's playgroups. So I decided to concentrate on children's play activities and toys in the first place (Rossie, 1993). Having elaborated a quite complete analysis of the Ghrib children's play activities and toys, I looked for information on Saharan and North African play, games and toys in the related literature and in the *Musée de l'Homme* in Paris where I found an important collection of toys from these regions. Since 1992 I have been conducting yearly research periods in Morocco, especially in rural areas and popular quarters of towns.[1]

Trained first as a social worker and then as an Africanist, my research methods belong first of all to the ethnographic research tradition based on participant observation, informal observation, informal talks, open interviews, use of informants and interpreters, making slides and doing some ethnographic filming. Moreover, I am using a detailed descriptive approach with a qualitative perspective when analyzing specific children's play activities and toys, and the sociocultural context in which these take place. Afterwards, the data of my own research, the information gathered from the relevant bibliography, and analysis of the toy collection in the *Musée de l'Homme* are used for a comparative analysis. Finally, I try to build a comprehensive description of the play activities and toys of the Saharan and North African children. Yet, this description should by no means be seen as a finished study. On the contrary, it is only when other scholars will verify and supplement my data and the interpretations I have elaborated, that a more objective and representative view can be worked out. Since Theo van Leeuwen introduced me to social semiotics in 1997, I try to apply this approach to my data.[2]

My play and toy research is available on the website of the Stockholm International Toy Research Centre (http://www.sitrec.kth.se - see Publications: Books/Articles), in the more general book, *Toys, Culture and Society: An Anthropological Approach with Reference to North Africa and the Sahara* (1999), and in the descriptive series, *Saharan and North African Ludic Heritages* along with *Children's Dolls and Doll Play* (1999). Also available at the website are *Commented Bibliography on Play, Games and Toys* (2003a), *The Animal in Play, Games and Toys* (2003b), *The Domestic Life in Play, Games and Toys* (2004, 130), and to be available with projected publication dates are *The Games of Skill* (2005) and *Traditional and Modern Techniques in Play, Games and Toys* (2006).[3]

MAJOR TOPICS OF CROSS-CULTURAL ANALYSIS

In addition to the descriptive approach, I am using my data to work out a cross-cultural analysis of the relationship between toys and play, on the one hand, and, on the other, child-child or adult-child relations, gender, rituals, festivities, creativity, change, and signs and meanings. Due to editorial limitations it is impossible to offer a detailed description and analysis of the afore-mentioned play activities and toys or to include photographs. I therefore refer the reader to my recent publications available on the Internet and showing many photographs and designs.

In Morocco my research has since 1992 and up to now been limited to central and southern regions. This vast area is composed of some northwestern plains, the mountain chains of the Moyen Atlas, Haut Atlas and Anti-Atlas, and the Pre-Sahara. All the children whose play activities and toys I have studied belong to sedentary communities and more or less popular families. With the exception of the two larger cities of Marrakech and Kenitra, these children live in villages and rural centers. Most of the information, especially that which comes from rural areas, refers to Berber-speaking children. But in other places and especially in towns, with the exception of Goulmima and Midelt, it concerns Arabic-speaking children.

Child-child and adult-child relations in play and toys

A few times I have seen Moroccan children playing alone, so it probably is not that exceptional. Yet, I have only observed such solitary play among small children as in the case of a three-year-old boy from the Moyen Atlas village Amellago playing in a small irrigation channel, making the little walls of his garage with mud bricks and using an old sandal as truck, or a five-year old girl from Midelt playing before her house door with wet sand and a plastic kitchen utensil. Still, children's play activities in these regions are mostly collective and outdoor activities. Playgroups are hereby the basic social organizations. After the age of seven years they consist of only girls or only boys, seldom of boys and girls together. When girls and boys form a playgroup together they are toddlers or somewhat older children, possibly under the direction of an older girl, eventually also an older boy. The factors for choosing playmates are primarily based on ties of kinship and neighborhood and this certainly strengthens the cohesion of the playgroup and the bonds between the children. Among older children schoolmates or other unrelated children can be part of the playgroup.

As far as I could observe this, the playgroups are organized by the children themselves and on their own initiative, but this does not mean that there are no leaders, no girls or boys who take more initiatives than others when play activities are concerned. When an older child plays with younger ones leadership comes naturally to the foreground, but when peers play together decisions on what to play and how to play are prevalent. Much information on how to play and make toys, but also on the environment, on cultural topics and on social relations, is observed or learned through playgroup interactions. It should also be stressed that these children create most play situations and toys to communicate with other children, whereas most toys and many games in Western communities are cultural messages created by adults for children.

In general, children's playgroups enjoy a lot of autonomy and freedom as in most cases adults do not take part in children's play, except when playing occasionally with a tot or a toddler. Intervention of adults in children's play seems rare and is mostly limited to insisting on taking care of little children, to intervene when play situations runs out of hand seriously, to ask for help from a child, especially a girl, or to react when the children's play disturbs an activity or a possession of an adult. So, most play and toy making activities I have found until now do not involve adults. Yet, I have seen or heard of Moroccan women and men playing with or making toys not only for a little child but also older ones. Some examples show a mother or a father making a toy-musical instrument or a toy-windmill especially for feasts. Artisans also make a small number of different toys that children receive at the same occasions. It also happens that a parent or adult sibling brings back from a trip to the weekly market a little present for a small child, often some cheap plastic toy imported from China. Nevertheless, the gift of toys so important in societies more oriented towards consumer goods remains exceptional. If it is not the child himself or herself who makes the toy, it can be his or her brother, sister or cousin. And even when adults make toys, these toys do not seem to fit into a system of rewards and tokens of affection.

The available information suggests that Moroccan children's pretend play, as exemplified in doll play, household play and play connected to the relationship between humans and animals, only shows locally valued situations and positive adult models the child identifies with. However, to see these children rigidly set to passive attitudes regarding the models of adult life would be erroneous. On the contrary, they appropriate, adapt and change these models according to their own needs.

Gender in play and toys

Moroccan toddlers are seen playing in heterogeneous as well as homogeneous groups. Yet, lacking any statistical data I would not dare to guess what is most common. A lot probably depends on the availability of playmates of the opposite sex. I also observed now and then young boys and girls of seven or eight years playing together. In three examples from 1999 the children use a small house. In Midelt two boys and a girl use a demolished van as small house and in a Moyen Atlas village near Amellago three girls and two boys between six and eight years play together in an elaborated small house delimited by stones and containing a large series of sun dried clay utensils. In a nearby village four girls and one boy of about seven years play at wedding in their doll's house.

In Moroccan children's playgroups and from the age of about six years onwards, gender differentiation becomes stronger. At that age, boys and girls create their own playgroups from which the other sex normally is excluded. Boys enjoy more freedom in their playgroups than girls in theirs, at least as long as norms are not too overtly transgressed. They also have the opportunity to go further away from their homes, the distance broadening as the boys become older as in the case of a Moroccan group of boys playing in the sea at two hours walking distance from their village. This way the boys can escape the direct control exerted by their parents or other adults. The girls on the contrary are not allowed to go far away alone. They must also stay near their mothers to help them in the household or because they have to look after some small children. When looking after these small children girls certainly do find occasions to play. Yet, the boundary between the task of amusing and occupying the little ones and the possibility to amuse oneself is difficult to draw.

Another clear difference between boys and girls is the time they have to play and this because of the girls' greater integration into household activities especially from the age of about eight years onwards. A striking example of this more limited time to play and more important integration of girls in household tasks is exemplified by an observation session of one and a half hour which I carried out in a small valley, serving among others as play area and situated between two popular quarters of Midelt. One summer morning in 1999 and during the time of the observation, I noted three playgroups made up by some boys and lasting between fifteen and thirty minutes. During the same observation time I found no girls playing, neither alone nor in groups. Instead, I saw one six-year-old girl cleaning the space before her house, another somewhat older girl passing by with a plate of biscuits on her head taking them to the oven and two girls doing errands. A fifth girl of about ten years was taking care of a group of little girls and boys.

As in other circumstances, one should always be careful generalizing statements such as the strict separation of older boys and girls in play because there are indications that this separation is not insurmountable, as the observation of play and toy making situations sometimes shows. Moreover, a few of my Moroccan female informants also stressed that as a child they liked to play with their brothers, cousins and other boys from the neighborhood, for example playing football or climbing trees. This makes it clear that the cultural norms of these regions are not the only determining factor in children's play activities and that the personality and the wishes of the players must be taken into account.

Rituals and feasts in play and toys

Some Moroccan children's play activities and toys refer to rituals and festivities, and some feasts are linked to specific games and toys. The girls' doll play offers a prominent example as it almost always represents wedding ceremonies. The representation is not limited to discussions for asking a girl in marriage, to dinner parties, singing and dancing. Specific rites strongly embedded in local beliefs are also acted out, rites such as the application of henna, replaced by wet sand, the mounting of the bride on a sheep to favor good life or the verification of the proof of the bride's virginity, a white rag on which red saffron has created blood-stains. The female doll invariably is called 'bride,' as well by the Arabic-speaking as the Berber-speaking girls, and the occasional male doll 'bridegroom.'

A burial rite can also arise all of a sudden in the imagination of Moroccan children, as I witnessed in a popular quarter of Kenitra. There I saw how a little child caught by four girls suddenly changes into a dead child that is transported at hands and feet, put on the ground and mourned by the girls shouting "Allah, Allah."

Looking for relationships between festivities, games and toys in Morocco, Ashura, celebrated at the beginning of the Islamic year, certainly is the most important festive period because it is customary to give sweets and presents to children. At that time the markets are overflowing with toys. Often they are cheap plastic toys. These last years, water pistols have been added to the musical toys, toy-beauty sets, toy-utensils, toy-vehicles and toy-weapons. However, there still are next to the plastic musical instruments also those made locally. The boys groups and girls groups use these musical instruments to accompany their singing when they walk from door to door to get sweets or some money. A drum with a skin tightened at both sides of a small cylinder is also given to the boys but an old plastic oil can replaces it adequately. The same oil can drum was used by Goulmima pupils for the Feast of the Dynasty in 1996.

An important ritual and long-standing playful aspect of Ashura is linked to the old water throwing custom. Children of earlier generations could permit themselves a lot of liberties when throwing water on children and adults. During the 1979 Ashura a group of girls and boys of about eleven years entered the mosque of Ksar Assaka near Midelt, took the pots filled with water used to perform the ablutions before praying and went on the flat roof. There they waited until someone passed by. A few minutes later, a man arrived with his mule loaded with a huge pack of herbs. The moment he passed before the mosque, the children threw all the water on him and his mule. As the man lost control over his mule, the pack of herbs fell on the ground. Yet, the man did not show hard feelings and the children came down from the roof to help him put all the herbs back in place. Those who told me this and similar anecdotes said that they thought that today the adults would not tolerate such behavior or that they would react angrily. In Midelt and during Ashura of April 2001 it became evident that water pistols and water guns have replaced former ways of throwing water.

Certainly in the regions of Goulmima and Midelt, the feast of the birthday of the Prophet Mohammed, for boys, a specific type of windmill characterizes the Mulud. In order to make the sail of their windmill turn quickly, the boys, and more rarely a girl, run with it very fast. Sometimes boys make these windmills to sell for about 1 *dirham* (0.1 Euro). In Midelt and some surrounding villages, I saw that the boys still make such windmills during the Mulud of June 2000. Yet, in one village I did not see these windmills and a few adults told me that the children of their village did not play with them anymore. However, in the afternoon of the same Mulud some children were running over the Midelt streets with a windmill bought from a hawker. One of these hawkers told me that an old man made them.

Two examples related to the Aïd el Kebir, the feast of the sacrifice, show how a mother and a father from the Midelt region make a little tambourine for their daughters. In the first instance and during the 1970s, a mother usually made a tambourine with the skin of a small sheep, sacrificed for this feast, to give it to her daughters. In the second instance and in 2000 a father made a little tambourine for his two and a half-year-old daughter. The girls use this tambourine when singing and dancing.

Swinging seems linked to the Aïd el Fitr, the feast at the end of the fasting month Ramadan. In a popular quarter of Kenitra I saw how children used the ropes attached to trees as swings for this purpose, perhaps referring in this way to old playful agricultural rituals in relation to the changing of seasons, just as in the case of traditional ball games.

Creativity in play and toys

Every toy made by a child and every play activity certainly is a creation, an original act resulting from the child's personality combined with the influences from the physical and human environment in which the child lives. A few years ago the Fourth Nordic Conference on Children's Play invited me to discuss this topic much more in detail (Rossie, 2001b). The play activities and toys mentioned here only represent some specific examples of Moroccan children's creativity.

My own view on the concept of creativity in relation to toys and play refers to the idea of making something, for example a toy, or working out a play, possibly a pretend play. I would also include the idea of doing something unusual, something original or even something classified as aesthetic and artistic. However, by including these ideas I am provoking a real difficulty and this because it will be necessary to oppose what is locally seen as unusual, original, aesthetic or artistic by insiders, the members of a given child's community, and what is externally seen as unusual, original, aesthetic or artistic by outsiders, the non-members of a given child's community.

Because of the primordial importance of playgroups, I want to put forward the hypothesis that Moroccan children's creativity in playing and making toys might more often be expressed, and if so should be investigated, in the children's interactions within their playgroups rather than in the case of isolated players. Moreover, If one is looking for purely individual solitary creativity among children, in the sense of a child making a toy or realizing a play event not only alone but also in a strictly original way, it will be difficult for me and probably also for other researchers to find instances of this among North African children even today. Moreover, I think that such a solitary individual creativity - although existing among Moroccan children as will be shown later on—is only one of the possible forms of creativity. After all, it could be that it is not so important or even impossible to know if and in how far a child has 'invented' a toy. Take for instance the creation by an eleven-year-old Central Moroccan boy of his own copy of a local musical instrument completely made with waste material. Although I have not seen another such toy-instrument, one cannot know if it is a personal invention of this boy except through the boy's own affirmation that it is so. Yet, even in this case it still is possible that others have more or less consciously influenced the boy. Nevertheless, if it is not this boy who has created such an instrument first, it is another one from his community who did it as in Morocco examples of self-made toys are not shown on television or in other media and, an eventual exception left aside, neither are proposed in schools or youth houses.

Children's inventiveness in the use of natural material of mineral, vegetal, animal and even human origin is omnipresent in North Africa. Four Moroccan examples, two of girls and two of boys, will illustrate this. At the same time some of these examples show how specific material is chosen to serve specific purposes. I have found a clever use of reed-leaves to create the hair of their dolls by the girls of a village near Midelt. To give their dolls the much-valued very long hair, these girls look for the upper part of a reed with long green leaves, leaves they split with their fingernails into small strips. A girl from a village near Khemisset uses stones not only to delimit her doll house but also transforms one of her three undistinguishable dolls into a bridegroom doll by putting a little stone serving as head in the upper part of the dress. Moroccan boys from a High Atlas village use summer squash, pieces of potatoes and sticks to make human and animal figurines, such as a mule and its driver. Moreover, I found on a road in southern Morocco a boy running with a self-made car for which he used two floaters of fishing net as wheels.

The children's creative use of material certainly is not limited to natural material as they also excel in re-utilizing waste material. A similar series of examples as for the use of natural material is offered here. When Moroccan girls play household they use whatever kind of waste material they can lay hands on and suiting their needs. Waste material is also used when girls make dolls as in the case of the bride doll mounting a toy-sheep. The making of this toy-sheep seems to be an original creation of the girl herself or at least of her playgroup as the sisters of that girl firmly stated that they never have made or seen among other girls such a toy-sheep. The toy-cars and toy-trucks of the Moroccan boys show the great variety of waste material used to make them, among others old oil filters and ball bearings for the wheels. Some years ago, I saw a thirteen-year-old herds boy sitting at the side of a road in the Middle Atlas while playing on a self-made violin. Members of the boy's family and some of his neighbors said that they knew of no other boy from the region doing the same. Yet, two men from Midelt told me that they also made a violin when they were older boys.

For a very common play activity Amal and Leila, two eight-year-old girls living in a Central Moroccan village in 1999, use really original toys. As Amal's mother forbids her daughter to play outside in the 'dirt,' this girl makes a dollhouse out of a cardboard box. Once the upper side of the box is cut of, windows and a door are made in the vertical sides in such a way that they can open and close. The interior side of the windows and the door are decorated with a curtain. There are a few self-made cushions and large or small rags serve as carpets and blankets. Leila has the same doll's house and together they play at the wedding of their bride doll. This bride doll is as special as is the doll's house. It is a Barbie-like imported plastic doll one can buy in local shops but that normally is only used as a decorative object once an older girl or a woman has crocheted an Andalusian dress for it. The two girls have both the same doll and they have sewed a dress for it. Looking more closely at Amal's doll one sees that the missing arms are replaced by a piece of reed, relying on the way traditional dolls are made.

Transferring new experiences to common toys is another way to be creative. A fine example of this creative process was shown to me in a village near Midelt. Until then, the boys made a truck with an oil can, four wheels cut out of a tire, a steering wheel of wire and so on. However, as they observed during the reconstruction of the irrigation system how a concrete mixer was filled with a lifting tray attached to the mixer, they invented a way to attach a lifting tray to their toy-truck using a small tin can tray and a long wire fixed to the steering wheel. When pulling the wire the sand or stones accumulated in the tray are thrown into the truck.

Self-made toys can be very simple but regularly they show a more complex elaboration, sometimes becoming a clever combination of many elements, as is the case of many dolls of the girls or of the toy-tractor of a High Atlas Mountain village boy, made with a piece of cactus as chassis and pieces of rubber as wheels and exhaust pipe.

One could wonder how it comes that Moroccan children from the 1990s and Tunisian Sahara children from the 1970s, living in non-industrial communities, playing games and making toys that reflect local situations, are so creative, with creativity being defined here as performing or creating something personally and independently from adult interference. In this context and looking at the way in which they grow up to become responsible members of their family and community, I would put forward the role of these children's own initiative in observing and in playing.

Change in play and toys

In Moroccan towns like Marrakech, Kenitra, Goulmima, Ifni, Khemisset and Midelt, imported plastic dolls have replaced the self-made doll. In Ifni, a small southern coastal town, girls still played about 1985 with self-made dolls having a frame of reed but nowadays the girls play with plastic dolls. But even if a six-year-old Ifni girl plays with a plastic doll, the other items used in her doll play remain unchanged. So, she places her plastic doll in a doll's house, the little square of paving stones before the door, and as utensils she uses a miniature wooden table with on top a few oil can stoppers filled with water and figuring cups of tea. Moreover, when making a first video on doll play in Sidi Ifni in January 2002 a young girl playing with her little brother used not only plastic dolls but also made herself the traditional dolls with a cross-shaped frame of reed (Rossie and Daoumani, 2003). In several Moroccan villages one finds today as well the self-made doll as the plastic doll, a plastic doll some-times adapted to local ways by giving it a self-made dress.

Toys made by the children themselves are often very short-lived play ob-jects. So they are remade again and again, this way offering possibilities for change through internal and external influences. Change, or maybe more correct progress, due to ameliorated skills because of exercise and the child's own development, whereby the toy becomes better adapted to the ludic functions it should have according to the child. Change because of environmental influ-ences, such as new material, learning from others how to do, shifts in interest promoted by social and economic change, influence from Western visual communication systems and global toy marketing.

Changes in toys and games do not mainly come from abroad, as in the case of Asian or European toys. On the contrary, it is interesting to notice that changes often occur in two ways: by using local material and techniques to create toys referring to new items, for example a clay telephone, and by using new material and techniques to produce toys referring to local themes, for example a cardboard box doll house.

Self-made toy-animals certainly do not have the function of teddies and other soft toy-animals. Still, such teddies and possibly also the affective relationship with this kind of toy-animals, are finding their way into popular households. At the end of 1994, I have noticed for the first time a teddy in a house in Midelt. This teddy was bought on a market with smuggled goods near the Spanish town of Melilla in the North of Morocco. It certainly was not intended for a baby or toddler but exposed on the television set as a decorative object. Nevertheless, a three-year-old girl stood four years later in the doorway of her house in Midelt holding a teddy in her arms and in 2002 a Sidi Ifni toddler walked around with his teddy and told me that it could not speak because it lacked a suggestion of a mouth.

Locally made or imported plastic toys, often of poor quality, can be bought in shops and markets. However, even when playing with plastic toys, the skills learned by making toys can still be important as in the case of a plastic toy-truck with a broken axle that a six-year-old Moroccan village boy repaired with a stick.

When analyzing the evolution of children's toys and games in North African countries, the influence of the emigrants must also be taken into account. When these emigrants return to visit their family they do not bring with them useful presents only but also prestige presents, among which may be dolls, toy-animals, toy-weapons, or bicycles.

The development of tourism also affects children's toys. Today in the east of Morocco, where tourists come to admire the sand dunes of Merzouga, some young girls make their traditional dolls with a frame of reed not so much any longer to play with them, although they still use them for their doll play, but for selling them to tourists. This way these dolls change from children's toys to touristy objects. The same evolution, but more likely referring to the toy-animals and toy-cars made by boys, can be observed in other Moroccan and African tourist places, possibly changing a child's play into child labor.

As the evolution towards a consumptive society is slowly but surely moving on, many children whose parents cannot afford to buy good quality toys and therefore often buy uncompleted, damaged or poor quality second hand toys, not only will feel frustrated but at the same time they become less motivated to make themselves the toys they usually play with. These cheap toys can be dangerous for children because safety control for toys is lacking in the region. This commercialization of toys also stimulates the attitude of looking at toys as gifts from adults to children, an attitude that until recently was as good as non-existent.

In general, one can claim that the self-made toys are quite quickly declining in the cities, a few exceptions left aside, such as toy-cars or toy-weapons made by boys. Moreover, the traditional self-made doll seems as good as forgotten in these cities. In any case, I only have found one Moroccan city girl still making dolls herself. Nevertheless, many children, largely but not exclusively in rural areas, still have much fun in creating their own toys. Yet, the availability of new material, for example plasticine that now can be bought in the little grocery shops, combined with the influence of schooling and television programs might stimulate a child to create something completely new such as a miniature toy-dinosaur made by an eight-year-old boy. Another example is the use of the packaging of a liquid that after freezing becomes a lollipop. This plastic packaging is about 19 cm long and 3.5 cm wide. Once the lollipop has been eaten, the child blows up the packaging, rolls it up starting with the open end, keeps it rolled up in his hand with the rolled part between thumb and index, and then suddenly releases the rolled part near the cheek of another child. If done by surprise and in the correct way, the viewed child jumps up and everybody starts to laugh. The fun of the game is to be able to do it by surprise as the children all keep this packaging with them.

The influence of the commercial amusement and toy industry brings about important changes in the play activities of especially teenage boys from urban areas. Even in small Moroccan towns one finds today several playrooms, organized in a garage or a small café, offering for some money games such as billiards, table football and pinball. Electronic toys are also finding their way into small towns like Midelt. There I witnessed in 1999 the craze of three twelve-year-old boys for a simple electronic toy with twelve game possibilities. This electronic toy had already been handed over between two or three friends before it came into the hands of the actual owner and it was certainly to be given to other boys of the peer group when the boy using it has tried it out. Shortly before writing this article an eleven-year-old boy told me of a remarkable and as far as I know unique example of the influence of television on the play behavior of Moroccan children from a popular milieu, namely the craze of Midelt's younger children for all that refers to Pokemon provoked by an Arabic spoken version of the Pokemon animation films broadcasted by the Moroccan television since the beginning of 2001.

It is clear that the ludic activities of the girls remain longer within the sphere of tradition than those of the boys who willingly find inspiration in technological innovations and sociocultural changes. But how to foresee the short-term and long-term influence on the girls of schooling and television that nowadays have found their way into isolated areas.

Signs and meanings in dolls and doll play

Although my analysis of signs and meanings in Saharan and North African children's toys and play activities directly is influenced by social semiotics, I do not use this scientific label anymore because my approach is limited to the descriptive level without developing the theoretical approach typical of social semiotics. According to Gunther Kress and Theo van Leeuwen, "Social semiotics is an attempt to describe and understand how people produce and communicate meaning in specific social settings. . . . Social semiotics is sign-making in society" (1996, p. 264). Adopting this point of view I shall discuss the materiality of Moroccan dolls, the related technology and some cognitive and emotional aspects of Moroccan children's dolls and doll play.

There is no doubt about the importance of materiality both in creating toys as in analyzing their signs and meanings. Yet, it is difficult to bestow meaning on the children's choices of the material they use to make toys, except the meaning of conformity with the ecological and sociocultural environment in which they live. However, can one stick to the idea that almost all the toys of the North African and Saharan children are made with non-durable material just by accident? Or is it not more so that at the basis of this fact lies the common practice of making a new toy whenever the children need one for their play activities. This practice certainly is fundamental as even when the toys easily do last for some time they only seldom are used again for a next play activity. Instead, they are often deliberately left behind or even destroyed, for the making of a new toy is part of the fun of the play activity.

The non-durability of self-made dolls contrasts with the greater durability of imported dolls, mostly plastic dolls. The few examples, I know of, that a Moroccan girl had an imported plastic doll, she had it for at least some time, possibly using it further on when a leg or arm was missing or when she had to give it a self-made dress to replace the original one. But can one conclude from the difference between the short lived, self-made traditional doll and the longer living plastic doll that for the girls themselves the plastic doll is more important? I don't think so, especially when looking at the play activity itself in which a traditional doll more adequately represents the bride, which is the central figure of most doll play. Nevertheless, the imported plastic doll is gaining importance through factors lying outside the girls' play activities, because: it is purchased and as such has a financial value; it is imported and thus belongs to the outside world; it still is a rare item in rural areas and among children from popular milieus and therefore brings prestige to those who have it and longing to those who don't have it. To make a doll oneself becomes a slow activity for poor, rural girls, "backward girls" they say in town, and is something urban girls do not want or should not do.

A few examples of the representational meaning of specific material when making dolls have been given when talking of the creative use of natural material. However, this intentional use of material and objects is not limited to making dolls. It is also important in the creation of other toys, for example when children use all kinds of round, cylindrical and oval objects to make wheels for their carts, bicycles, cars, trucks and tractors.

When making their own toys the Moroccan children are restricted to what is called the "technologies of the hand." The hand tools are mostly objects they find themselves, not tools of adults, objects such as stones or other heavy objects to hit, the child's own teeth or other sharp objects to cut or make holes. One technological aspect to be solved by toy making children is movement, movement of parts of the toy or movement of the whole toy. Some toys such as windmills, toy-vehicles and toy-weapons have movable parts. In contrast with the imported dolls, the self-made dolls I know of have no movable parts. Yet, the fact that they are not articulated should not be attributed to a lack of technical know-how as other toys have movable parts. So the girls could have given movable arms or legs to their dolls if they wanted. A simple explanation for this situation would possibly stress the fact that in the eyes of the girls there is no necessity to do this as they themselves are assuring the mobility of the doll through their manipulation of it and because it is a very short-lived doll. An ideological explanation might be found in the argument that a doll with moving arms and legs is more like a human being than a rigid doll, this way possibly falling more directly under the Islamic prohibition of creating images of living beings.

The movement of the rigid doll is under the direct control of the child who manipulates it. The doll's movements are not naturalistic but conventional and based on a simplification of reality. What is important is the meaningfulness of the movements not their realism. Three sisters from Ksar Assaka, a village close to Midelt, explained that they and the other girls of their playgroups moved a doll by holding it at the lower end of the reed, making with the doll held upright back and forward, left to right and up and down movements. The doll was also twisted around especially while singing and imitating the wedding dances. When moving the doll this was clearly done at eye level, what according to Gunther Kress and Theo van Leeuwen reflects a relation of equality between the bride doll and the playing girls. An argument for the plausibility of this interpretation can be found in the fact that when the same girls used another doll for a ritual to obtain rain, the special status of this representational figure, once a North African female deity, became visible because the girl wearing this doll held it high up above her head while walking around the village.

The self-made dolls of the region can be qualified as analytical structures rather than naturalistic ones. They have been designed to show significant attributes and characteristics of the model they represent. Their makers are not interested in representing an individual living example of that model but in making a symbolic representation of a sociocultural role. Yet, the self-made doll itself, as bearer of individual and social meanings, is nevertheless treated with a lot of indifference once the play activity is over. Could this be the reason why an individual name for a doll was almost never mentioned to me by my informants from Morocco or the Tunisian Sahara or that not one bibliographical document has mentioned an individual name for a traditional doll?

The girls' affective relation seems to be directed towards the representational concept, the represented model, rather than to the doll, the material realization of the concept or model that is used as a means and only valuable as long as the play activity goes on. One might say that the function of such a doll is limited to the game; it only comes to 'life' when the player manipulates it, when it becomes part of a series of interactive relations mutually accepted and enacted by the members of the playgroup. When the play activity is interrupted or stopped, the doll becomes an object, a material item that can be left on the spot or thrown away. It certainly does not become the substitute companion doll Brian Sutton-Smith (1986) describes in relation to recent North American childhood.

When commenting on my analysis, Theo van Leeuwen stressed that the way in which my research provides great examples of semiotic recontextualization or re-interpretation is semiotically interesting, for example when Western plastic dolls become Moroccan brides. Two of these examples of recontextualization or re-interpretation are related to imported plastic dolls. The first example is a cheap imitation of a Barbie doll used by the above-mentioned girl who created a cardboard dollhouse. The second example is located in a really poor quarter of Marrakech where most of the girls still played about 1980 with the traditional self-made doll having a frame of reed. But a girl living in the same quarter then already played with an imported doll. This girl, now a woman skilled in the embellishment of hands and feet with henna-designs, transformed a cheap plastic doll into a wonderful bride of Marrakech.

Play activities and toys are strong signs pointing to the interests of the children. At this level, a striking difference between girls and boys is found, the girls being often engaged in doll play, household play and dinner party play, and in making the toys related to these games, whereas the boys are much more engaged in play related to transport and technology and in games of skill (Rossie, 2001a), and in making the toys used in these games.

NOTES

1. From 1975 - 1992 my research has been subventioned by the Belgian National Foundation for Scientific Research, Brussels.
2. Next to a scientific approach, I also use my information for some pedagogical and sociocultural activities to promote cross-cultural understanding and peace. A description of these activities is given in Rossie, 2003 (see Perspectives) and in Rossie, 2001b.
3. The NCFL center closed in 2001, and in 2002 was reestablished as the Stockholm International Toy Research Centre at The Royal Institute of Technology (SITREC). One should consult the publications formerly on NCFL's website on SITREC's website: http://www.sitrec.kth.se. When problems arise in consulting these publications please contact the author by e-mail: jprossie@hotmail.com.

REFERENCES

Kress, G. & van Leeuwen, T. (1996). *Reading images: The grammar of visual design.* London: Routledge.

Rossie, J. P. (1993). Children's play, generations and gender with special reference to the Ghrib (Tunisian Sahara). In C. Gougoulis (Ed.), *Special Issue on Children's Play,* (193-201). Ethnographica IX, Peloponnesian Folklore Foundation, Athens.

Rossie, J. P. (1999). Children's dolls and doll play. In *Saharan and North African ludic heritages.* Second edition, SITREC-Stockholm International Toy Research Centre, Stockholm, 116 ill. Available on the website http://www.sitrec.kth.se, see Publications.

Rossie, J. P. (2001a). Games of physical skill from the Tunisian Sahara and Morocco: Anthropological Research and Physical Education for Peace, *Actas del I Congreso Estatal de Actividades Físicas Cooperativas, Medina del Campo:* July 9-12, 2001. La Peonza Publicaciones, Valladolid, 64 ill., published on CD-ROM, ISBN 84-921266-3-9, Internet URL: http://www.terra.es/personal4/lapeonza/
e-mail: cvelazqu@roble.pntic.mec.es

Rossie, J. P. (2001b). *Children's creativity in toys and play: Examples from Morocco, the Tunesian Sahara and peace education.* Fourth Nordic Conference on Children's Play, Hämeenlinna, 81 ill. Available on the website http://www.sitrec.kth.se, see Publications.

Rossie, J. P. (2003a). *Commented bibliography on play, games and toys.* SITREC-Stockholm International Toy Research Centre, Stockholm. Available on the website http://www.sitrec.kth.se, see Publications.

Rossie, J. P. (2003b). The animal in play, games and toys. In *Saharan and North African ludic heritages.* SITREC-Stockholm International Toy Research Centre, Stockholm, 103 ill. Available on the website http://www.sitrec.kth.se, see Publications.

Rossie, J. P. (2004). *The domestic life in play, games and toys.* SITREC-Stockholm International Toy Research Centre, Stockholm. Available on the website http://www.sitrec.kth.se, see Publications.

Rossie, J. P. & Daoumani, B. (2003). *Protocol of video 1: Doll play and construction play in Sidi Ifni, Morocco.* January 31. SITREC-Stockholm International Toy Research Centre, Stockholm—Video and protocol placed in the video library of SITREC.

Sutton-Smith, B. (1986). *Toys as culture.* London: Gardner Press Inc.

Chapter 6

At Work At Play: Training Hindu Nationalist Women

Kalyani Devaki Menon

Training camps organised by the New Delhi wing of the Rashtra Sevika Samiti (henceforth Samiti), a women's organization that is part of the right-wing Hindu nationalist movement in India, provide a context in which to explore "play" as a locus for the dissemination, as well as the contestation, of nationalist ideology.[1] These training camps typically include two types of activities: the *baudhik* sessions that focus on intellectual development, and the __r_rik sessions that focus on the enhancement of physical strength and dexterity. Both these sessions encompass activities that would fall under the rubric of "play." The *baudhik* sessions intersperse periods of lecture and discussion with storytelling sessions, the learning of prayers, patriotic and devotional songs, as well as Hindu nationalist songs written by the ideologues of the movement. The __r_rik sessions include the study of military commands, rigorous exercise, endurance training, yoga and games.[2] Play becomes a critical site for ideological dissemination, one, as I will show, which can be far more effective than the lectures and discussions particularly for the youngsters who were the main participants in this training camp. It also provides the space for challenge and is thus constantly subject to supervision, manipulation and control. The ideological power of play in these Samiti training camps rests in its ambiguity; however, it is precisely this quality that permits the participants to challenge the movement and undermine its ideological prescriptions.

While the focus of this paper is on "play," the intellectual sessions provide the socio-cultural and political context in which this "play" must be understood. The most direct efforts at disseminating the ideology of the Hindu nationalist movement to the new recruits came during the lecture and discussion periods of the *baudhik* sessions. During these periods various leaders of the Samiti would lecture to the participants about the movement's ideology. Central to these lectures are themes that are critical to the Hindu nationalist movement: the idea that Hindu culture and society is under siege by Muslims and Christians who seek to convert them; and the idea that the new values, whether Christian/Western or Muslim, are inherently corrupt and threaten the honour and sanctity of Hindu society. Key to all the lectures delivered during the camp, is the idea that the

Hindu nation, its culture, and its religion are being threatened by those who are Christians and Muslims. Although Christianity in India can be traced back to 52AD and Islam can be traced back to 629AD, both these religions are considered foreign by the movement. As in much Hindu nationalist discourse, Christianity is seen as embodying the values of the West: sexual freedom, moral corruption, and imperialism. Islam too is viewed as corrupt, and, once again sexuality becomes a key factor in establishing moral decrepitude. According to members of the movement, Islam's corruption is "proven" because Muslim men can have four wives[3]. The lectures emphasize that it is only by creating a strong and organised force in society, as exemplified in the Rashtra Sevika Samiti, that these influences can be countered and the Hindu nation can be protected. This organization needs to have a wide base and have regular meetings that spread their message of the need to fight for, and indeed save, the Hindu nation. These messages are also encoded in the play sessions organized by the movement.

Victor Turner has argued that 'play' should not necessarily be seen as separate from and/or opposed to 'work' (a distinction born in the ideological shifts of the industrial revolution), but rather that work is often "an element of 'play'" (1974, p. 91). Brian Sutton-Smith asserts that while play has a life and meaning of its own, we must not abstract "the game from its play group context. Any game requires a gaming society, and any society has norms and hierarchies that interpenetrate the game" (1997, p. 106). What gives meaning to the play talked about in this paper is not simply the Hindu nationalist ideologies that penetrate it, but critically, also the energy of the play itself conveyed in the performative power of storytelling and singing, or the fun and excitement of the game being played. Building on Raymond Williams' concept of emergent culture, Richard Bauman asserts, "Performance, I would offer, constitutes ... the nexus of tradition, practice, and emergence in verbal art" (Bauman 1977, p. 48). It is, in other words, in performance that new meanings are linked up with older symbols and ideas.

Play is a performance in which all participants are brought together regardless of the ideological, class, ethnic, and other distinctions that might exist as part of everyday life. This does not mean that these differences are erased in this moment of communitas (Turner 1969), but rather that what brings the participants together as a community is prioritised. Arjun Appadurai asserts that in Victorian England although the game of cricket was able to bring together people from all backgrounds as a team, it is important to remember that "no amount of shared cricket would make an Englishman confuse an Oxford Blue with a Yorkshire working-class professional cricketer" (1997, p. 92). Similarly, in the case of the Samiti training camps, the ideological differences between the various participants, while suppressed, were not completely erased. The ability to render these differences unimportant, even for just that moment of play, is powerful enough to permit the realization of one's membership in a larger community, a symbol that resolves social differences, however momentarily, and has the potential for transformation into a "switchpoint of social action" (Turner, 1975, p. 80).

Complicating Victor Turner's contention that ritual symbols have transformative power (Turner, 1969), Mary Hancock argues that ritual is a key site "for the exercise of hegemonic power" in which "gender, class, and caste hierarchies might be reproduced" (1999, p. 19). However, it is important to remember that it is both "a site of force and resistance; in terms of subject formation, it creates complex subjects who are both compliant and resistant, subjects who can and do intervene in the streams of action that constitute ritual" (Hancock 1999, p. 19). James Brow argues: "the struggle for hegemony [must be] understood as the process whereby the interests of other groups are coordinated with those of a dominant or potentially dominant group" (1996, p. 24). I analyse play as a key site in the struggle for hegemony between the members of the Samiti and the participants in its training camps. At the Rashtra Sevika Samiti training camps, the meaning of play is similarly flexible and play can become meaningful to the audience for different reasons from simply having fun playing the games to enjoying singing a song. In contrast to the lectures, wherein the participants are clearly presented with the ideological positions of the movement and there is little room for argument or misunderstanding, the ambiguous nature of play (Sutton-Smith, 1997) allows a space for hegemonic struggle in which the interests of the participants (having fun, winning the game, enjoying a story, showing skill at being able to memorize a song and sing it well) are powerfully articulated with those of Hindu nationalism (expanding its base, making women physically and mentally strong, devaluing westernization, vilifying Muslims, and glorifying Hindu nationalist heroes).

Play, particularly in the form of stories and songs, becomes a key site for the dissemination of the values and ideological positions of the movement, which socialized young girls and women into adult members of the Hindu nationalist movement, thus becoming a "pathway over which ... [to] make a smooth transition to adulthood" (Loy & Hesketh, 1995, p. 98). Yet, as Brain Sutton-Smith argues, one should not merely reduce play to a tool for the progression from childhood to adulthood, because there are many similarities between the play of children and the play of adults (1995, pp. 280-281). This becomes very apparent in the Samiti training camps where children and adults, as in the case below, play together and listen to similar stories and learn similar songs. Importantly, one cannot assume that participants unequivocally accepted the messages coded in play. In fact, participants not only used these sessions to challenge the ideology of the movement, but also used them to subvert the type of community that the camp was trying to construct by defining themselves in opposition to the leaders of the camp.

Appadurai (1997) contends that cricket became a means through which the British conveyed the values and ideology of the Victorian elite, not only to the English working classes, but also to their colonial subjects. While the game provided a space for fashioning the colonial subject in the delicacies of sportsmanship, fair play, team spirit, and emotional control (Appadurai, 1997, p. 92), it also became the ground of contestation that ultimately overturned the Victorian values that it embodied (Appadurai, 1997, p. 107). Appadurai argues that today cricket is "No more an instrument for socializing black and brown men into the public etiquette of empire, it is now an instrument for mobilising national sentiment in the service of transnational spectacles and commoditization (1997, p. 109). This argument suggests that rather than simply being an ideological instrument that reaffirms cultural values and binds communities together, many forms of play (as often happens in the case of carnivals) can in fact "subvert" or "invert" the "typical conventions and decencies that are thought to bind that society together" (Sutton-Smith, 2001b, p. 44). The games I describe in this article are often very rough, aggressive, and competitive games in which the participants often take pleasure in pushing and shoving their opponents during the excitement of the game. While competition may be a value that the movement is trying to instil, the behaviour that results from this competition is sometimes antithetical to the kind of society that the movement envisions: one marked by order, obedience, discipline (particularly disciplined bodies), and authoritarian hierarchy (see Alter, 1994).

THE RASHTRA SEVIKA SAMITI

I conducted fieldwork in New Delhi between January 1999 and January 2000 with women belonging to the various wings of the Hindu nationalist movement in India. The Hindu nationalist movement has existed in various forms for over a century, but at no time has it been more powerful in the sociopolitical landscape of India than it is today. Multiple organizations are eclipsed under the rubric of Hindu nationalism, from the exclusive all-male paramilitary organization, the Rashtriya Swayamsevak Sangh, to the more inclusive electoral party, the Bharatiya Janata Party, that is currently heading the coalition that forms the national government in India today. What brings all these organizations together is the common desire to purge the country of all "foreign" (read: Muslim and Christian) influences and establish India as a Hindu nation.

There are several women's wings in the Hindu nationalist movement of which the Rashtra Sevika Samiti is only one. The Samiti's goal is to make women physically and mentally strong. This differs from the goals of other women's organizations in the movement such as Durga Vahini, which, as Payal Gupta a member of the Samiti pointed out in a deprecating tone, is "more interested in *dharm k_ k_m*" (religious work). Often referred to by scholars and journalists as the "women's wing" of the Rastriya Swayamsevak Sangh (RSS), the Samiti defines itself, in the words of Aparna Sharma, a leader of the Delhi Samiti, as "a parallel organization of the RSS." As other women took pains to tell me, the Samiti is one of the many "branches" of the RSS, in the same way that the Vishwa Hindu Parishad, Durga Vahini, Sewa Bharati and so on are branches of the same tree. This arboreal metaphor was employed time and again by both men and women of the various wings of the movement to describe the relationship of the various wings of the movement with the RSS. In this metaphor the RSS formed the trunk while the various wings, including the electoral party the

Bharatiya Janata Party, were the branches spreading out of the trunk in different directions, or with different specialities as it were.

The Samiti was founded in 1936 by Lakshmibai Kelkar. Neera Gupta, one of the leaders of the Delhi wing of the Samiti, described the Samiti's goals in the following way: "the Samiti's goal is to organise Hindu women and teach them to be mentally and physically strong and to teach them about their own culture so that they can pass this on to their children." A leading figure of the Samiti, whom I shall refer to as *Taiji*, or aunt, as she is called by Samiti members, explained that the Samiti had been started by Laksmibai Kelkar because she felt that Indian women had become too weak and frightened and she wanted to make them mentally and physically strong. *Taiji* asked those present in the room where she was addressing us "why had women become so scared?" Her own answer claimed:

> Ever since the Muslim invaders came to India there have been atrocities committed against women and, consequently, society was no longer a safe place for them. In ancient India Hindu society gave women a lot of freedom and there was no *pard_* (veiling) and women could wander around alone with no fear. All this has changed with Muslim invasions and now women have to stay at home, maintain *pard_*. Because in those days if the Muslims saw women and liked the way they looked they would just pick them up and force them to become part of the harem. Ever since these times, women have been unsafe and even today a woman out of doors is not safe. This is why the Samiti teaches women to defend themselves and also gives them mental strength to overcome their fear. To this end, weekly *_kh_s* are held in various parts of Delhi (25.11.99).

The Samiti organises weekly *_kh_s*, or training sessions, for women and girls that entail both physical and mental training. Physical training involves teaching participants yoga, martial arts, and how to fight with *l_h_s* (bamboo staves) and daggers. Mental training includes lessons about Indian history, politics, and culture. Aparna Sharma, one of the leaders of the Delhi Samiti, informed me that in the *_kh_s* for women, as opposed to those for younger girls, there is an emphasis on mental training, particularly "on discussions of current topics, women's issues, like dowry problems, and history." In addition to these *_kh_s*, the Samiti also organises *_ivirs* or training camps like the one that is the focus of this paper. These are intensive residential camps held over a period of several days (from three days to two weeks) that train girls and women to run the weekly *_kh_s*. These *_ivirs* follow a strict timetable, repeated daily, which divides the day up into *baudhik* and *_r_rik* sessions. To provide a sense of the intensity of these camps, Table 6.1 shows the timetable given to us during the three-day training camp that I attended:

Table 6.1

Training Camp Daily Timetable

J_gran (wake up)	5 am
Pratah Smaran (morning prayer)	5:45 am
Yog_san (yoga)	5:45-6:15 am
*Pratah __kh_* (morning __kh_)	6:15-8:00 am
Na_t_ aur sn_n (breakfast and bath)	8:00-9.30 am
Pr_rthan_, g_t abhy_s (prayer and song lesson)	9.30-10.30 am
Break	10.30-10.45 am
Baudhik (intellectual session)	10.45 am-12.30 pm
Bhojan (lunch)	12.30-1.30 pm
*Vi_r_nt_* (rest)	1.30-2.30 pm
*Charch_* (discussion)	2.30-3.45 pm
Tea	3.45-4.30 pm
*S_yam __kh_* (evening __kh_)	4.30-6.30 pm
S_yam smaran aur bhajan (evening prayer and songs)	6.45-7.45 pm
Bhojan (dinner)	8:00-9:00 pm
Manoranjak k_ryakram (entertainment program)	9:00-10:00 pm
Vande m_taram (hail to motherland, a patriotic song)	10:00 pm
D_p visarjan (lights out)	10.15 pm

Although everyone present knew that I was a researcher, I was also expected to be a full participant in the training camp. As a participant of the camp,
I can say with some authority that the program was exhausting and by the end of
the camp, I was both physically and mentally exhausted. Every minute of the
camp, except when asleep, we were being supervised, controlled and were constantly in performance, playing our part in the camp.

The Participants

I was informed about the _ivir by Payal, a twenty-nine year old *prach_rik_*
of the Rashtra Sevika Samiti. A *prach_rik_* is a woman who has devoted her life
to the movement, in contrast to married women who are *grihast_s* and have to
divide their loyalties and time between their families and the movement. According to Payal, there are only thirty-three Samiti *pracharikas* in the entire
country. There are many more *prach_raks* (or men who have devoted their lives
to the movement). *Prach_raks* and *prach_rik_s* are volunteers who, for the most
part, are never paid by the movement. Payal become a *prach_rik_* with parental
consent; however, there were other girls who came against their parents' wishes

and are given a small stipend to buy necessities. Payal said that very few women are able or willing to become *prach_rik_s*. She asserted that, in contrast, the all male RSS has thousands of *prach_raks* because it is less difficult for a man to reject *grihast_* (household duties) and devote his life to the cause. Having managed to get her parents to consent to her joining the Samiti as a *prach_rik_*, Payal left her family in Punjab and came to Delhi where she has lived and worked ever since. Payal is in charge of supervising the running of *__kh_s* throughout Delhi, a job that she loves. She told me that although many have told her to join politics because she now has such a large base of supporters, she prefers to do this work for the movement. Payal was in charge of both organising and running the *_ivir;* and although she woke up the earliest and went to bed the latest of all of us, still managed to exude energy and enthusiasm throughout the *_ivir*.

The *_ivir* was held in the compound of a Hindu nationalist school in South Delhi, then closed for the winter holidays. The school compound itself was situated within a larger compound that held the offices for various wings of the movement, as well as some limited residential facilities. We were not permitted to leave the school compound, and, nobody, except for the various ideologues who came to talk to us, was allowed into the compound for the three days of the camp. Immediately inside the gate of the school was an open area about the size of a tennis court, surrounded on three sides by the two-storey school building. Most of the camp activities described in this paper took place in this area or in the three classrooms that had been converted into living/learning areas situated just off the open area. Two classrooms were converted into dormitory style sleeping areas in which all the participants slept on thin cotton mattresses provided by the Samiti. The third classroom was converted into the *baudhik* room in which we gathered to listen to the lectures as well as to pray, learn songs, and listen to stories. The *__kh_s* were held in the compound outside.

Working directly under Payal were two *_ik_ik_s* (teachers), Ruchika, a young woman in her early twenties, and Maya, a fifteen-year-old girl. Both Ruchika and Maya had attended a prior fifteen-day training camp and thus were already eligible to be teachers. Maya ran a weekly *__kh_* in South Delhi. Ruchika did not run her own *__kh_* but liked being involved in the movement. Attending the camp had not been easy for her. Her parents had not wanted her to go and she had to fight with them to get permission to attend it. As it turned out, this fight was almost in vain. Payal had told us to arrive at four o'clock. I arrived at the school where the camp was to be held at precisely four to find nobody there. In fact the armed guard at the gate, as well as the people in the head office, did not even know that there was a camp to be held there that weekend. About forty minutes later, Ruchika, whom I had met once before, arrived with her father. She was very perturbed to find nobody there. Her father too seemed very unhappy about the situation, and Ruchika later told me that he had wanted to take her back home. Luckily Payal arrived at that moment and confirmed that there was indeed a camp and that the other participants would be arriving shortly. Payal asked Ruchika to tell her father to leave. He did but did not look too happy about leaving his daughter with two other young women and Ruchika later confirmed this. That night, after pleading with Payal for permission to leave the school compound, I accompanied Ruchika to the nearby public telephone so that she could reassure her parents that the others had arrived. She was still worried that her mother would send her father back to pick her up.

At about six o'clock, a tempo (a small open back truck) arrived full of people from various parts of South Delhi. Maya arrived on the tempo with her two sisters and her mother, all of whom are part of the movement. Maya was in the tenth grade and had brought her best friend Gauri along to the camp, as well as another friend Nita. Maya's younger sister Radhika, who was in the seventh grade, had also brought her best friend along, a girl called Leela. The youngest sister was only three years old and was basically there because her mother was. The girls were all very friendly and full of life. They were very curious about me and came and asked me all sorts of questions including some about my work. Gauri, who had the most questions for me, said she agreed to attend the camp only because Maya had asked her to. The first night after the lights were turned out the girls stayed up chatting about Salman Khan and Sharukh Khan, two leading male actors in the Hindi film industry, and boys. Gauri was teasing Maya saying that Maya did not want to consider Gauri as a sister because she had a crush on Gauri's brother and if they became sisters then he would become a brother to her. While obviously this is normal conversation for teenage girls, it would have drawn the ire of members of the elder members of the Samiti had they been there to overhear it because in the movement female sexuality is only considered appropriate within the sanctified space of marriage (Menon: nd).

Also in the tempo was a young woman called Anshu, doing the final year of her Bachelors degree in Commerce by correspondence, and two girls, Divya and Bina, in their first year at Aurobindo College in New Delhi. Divya's mother, a long time member of the Samiti, was also at the camp, but she along with all the other older women were basically there in a supportive capacity. They were responsible for cooking and making sure everyone was well fed and generally keeping a maternal eye on things. They attended the lectures and discussions but did not take part in anything else. Divya and Bina were also "best friends" and quite consciously separated themselves from the rest of the group. While everyone else wore *salv_r kam_z*,[4] they wore trousers with shirts and were criticised by Payal for not wearing Indian clothes. Both were consistently late for everything and were regularly reprimanded either for being tardy or for talking to each other when they were supposed to be paying attention to something else. It was only during game time that they really became part of the group and got involved and enjoyed themselves. Bina told me that she only came along for the _ivir because Divya was coming. She also said that she did not like all the things that were being said about Muslims and Christians and wished that they only had the games and not the *baudhik* sessions. She and Divya basically stuck to each other the whole weekend except during playtime. Bina and I were among the few who were consistently "bad" at learning the various Sanskrit _lok_s (verses) and songs that we were expected to memorize. She, like me, was basically mouthing words of songs we were supposed to have learnt – pretending to know the words. When we played *ant_k_in_*[5] of patriotic songs that first night neither she nor Divya had anything to contribute. The other girls were quite taken with Divya and Bina. When the news broke that Bina rode a scooter around Delhi, all the girls began to whisper about this to one another. The freedom to ride a scooter unaccompanied around Delhi was not something that most of these girls, from middle-caste/middle-class backgrounds, had.

In all there were twenty-one of us, not including the returning members (that is the five older women, Payal, Maya and Ruchika). The ages ranged from twelve to about thirty. One of the participants was a woman, Rajani, who was about thirty, had four children, the youngest of whom, a little boy, had come with her. I was informed on the first day by an RSS *prach_rak* who had arrived to help the women set up camp that two doctoral students from Jawaharlal Nehru University in New Delhi were also supposed to attend the camp. In the end these people never actually showed up.

At Play

While a major part of this section examines the games played by the Samiti during its training camps, play here also includes the song learning sessions, the folktale sessions, the general "rest" time in between sessions, as well as the entertainment program scheduled for the evening. I begin by discussing the entertainment program, the stories and the songs, and then go on to analyse the games played at the camp.

To demonstrate that play was indeed being strictly supervised and circumscribed, I begin by looking at the entertainment program during which we were supposedly allowed to engage in any group activity to entertain ourselves in the evening before bedtime. The first day, under the strict supervision of Payal and the older women, we played *ant_k_in_* (see Note 5) of patriotic songs. On the second day, Payal and the other older women were busy entertaining Rukmini, an older *prach_rik_*, who had just arrived, and Maya and Ruchika were put in charge of the program. Ruchika suggested we play *ant_k_in_* of patriotic songs again, but there were several protests from girls in the room. While they would never have voiced such protest in front of Payal, they had a different relationship with Maya and, by extension, Ruchika, and therefore felt free to challenge their authority. Someone suggested we play *ant_k_in_* of film songs. Both Maya and Ruchika said we could not do this because Hindi film songs did not have the right moral tone for the camp, but we could play it with patriotic songs. The other girls, including Maya's two best friends, got upset and began to complain until finally Maya relented and compromised on playing charades with Hindi movie titles. While everyone was having fun, Payal came in and the room fell silent. With a raised voice she told us that we should stop playing this game immediately. She said that we were at the camp to learn the proper cultural values and morals, not those taught in Hindi films. Then she called Maya and Ruchika outside with Ruchika proclaiming that she had been against this all along and it had not been her idea. Later that night Ruchika told me that Payal had scolded them for allowing things to get out of hand.

While it could be argued that the entertainment program organised by the Samiti would undoubtedly carry with it certain prescriptions, even the "free time" we had between sessions was strictly monitored and used by the Samiti to spread its message. On our first morning (the second day), Payal, Ruchika, Gauri, Maya, Radhika, Leela, Nita and myself sat around in our room and chatted after breakfast while the others were bathing and getting dressed. Payal turned the conversation to Valentine's Day[6]. Payal said that the real meaning of Valentine's Day was friendship but now society had become so *gand_* (dirty) that it has become all about boys and girls and love. Gauri said that she ties friendship bands on all her friends on Valentine's Day. Payal then began to talk about how Valentine's Day was just a western festival that had been brought to India. She said that it made no sense that there should be one day in the year when friendship is recognised. What about the rest of the year? She asserted that we should be wary of these westernisms in the same way that we need to be wary of the missionaries coming from the west and converting Hindus.

In this example, Payal turned a light-hearted discussion about who was friends with whom to one in which a key concept in Hindu nationalist discourse could be conveyed. In this discussion, Payal does not initially criticise Valentine's Day, but rather simply suggests that it has been misconstrued. When others join in the discussion, she introduces the idea that Valentine's Day is not part of Indian/Hindu culture but rather a festival imported from the west embodying the misguided notion that friendship only needs to be recognised one day in the year. She then caps the discussion with an idea that came up time and again throughout the camp, namely that Christian missionaries and western culture have the same agenda: to destroy Indian/Hindu culture by introducing these western habits and traditions so that people forget their own heritage.

The storytelling and song learning sessions were also integral forms of play at the training camps. Relating folklore about Hindu heroes is a key way in which the Samiti conveys its ideology in *_ivirs*. Folklore is a powerful means through which cultural ideals are translated and disseminated. Kirin Narayan contends: "A story is a 'cognitive instrument,' a means of making sense of the world" (1989, p. 100) and as such it is "an expression of deep-rooted cultural themes" (1989, p. 99). Folklore can often provide the locus for the embodiment and construction of historical memory (Gold & Gujjar, 2002). Nationalist folklore is a powerful means through which subjectivity is constructed for those belonging to the nation, both through the ideals embodied in the stories and through the insertion of individuals within larger historical trajectories, or "mythico-histories" to borrow a term from Liisa Malkki (1995). Malkki defines mythico-history as "a process of world making" concerned with "the ordering and re-ordering of social and political categories, with the defining of self in distinction to the other, with good and evil" (1995, p. 55) in order to constitute "a *moral order* of the world" (1995, pp. 55-56 original emphasis).

The _ivir_ provided several such opportunities for storytelling. In one instance, a Hindu nationalist *S_dhav_* (female ascetic) came in to talk to us and told us a story about Akbar, one of the greatest of the Mughal emperors well known for his tolerance for all faiths and his desire to reconcile Hindus and Muslims in his empire through the creation of his own syncretic religion. This story about Akbar's Meena *baz_r* (market) was related by the *S_dhav_* in the context of a longer lecture about the place of Hinduism and Hindu womanhood:

> This was a *baz_r* only for women. A rule made by Akbar. He used to dress himself and his courtiers as women and go and then try to trick the women into going with him. Once while this was going on Rana Pratap's niece was there and Akbar managed to get her to follow him. However, she was too clever for him. Kiran *m__* attacked him with a dagger like Durga with one foot on his chest. Akbar begged for forgiveness calling her '*m_*' (mother). She forgave him but made him promise to stop the Meena *baz_r*. Character is everything. Girls should aspire to be Savitri and Ansuya and Sita. Whatever advertisements you see on TV uses women's bodies to sell their product [26.12.99].

The *S_dhav_* concluded her talk saying; "Do you like it if someone comes and curses your mother? We should feel the same way about *Bh_rat M_t_* (mother India)[7] and be ready to fight for her. These _ivirs_ are very important because they are for all Hindus and cement society together" [26.12.99].

At an obvious level this story about Akbar vilifies the Muslim emperor as a lustful and immoral king preying upon his innocent Hindu subjects, a theme that is one of the organising principles of Hindu nationalist accounts of the relationship between Hindus and Muslims throughout history. Equally importantly, however, is the statement that it makes about Hindu womanhood. Particularly in terms of the agenda of the Samiti, it is able to construct an image of the strong and virtuous Hindu woman in the figure of Kiran *m__* who is able to defend her self and her virtue from defilement by the lustful king. Key to this discussion is the construction of motherhood. Kiran *m__* forgives Akbar when he refers to her as "mother," reflecting an idea held by many in the movement, that forgiveness is one of the key virtues of motherhood. Also, the exemplary womanhood demonstrated by Kiran *m__* makes her like the Goddess Durga, an analogy that reflects the movement's assertion that women, as mothers, who are pure, chaste, self-sacrificing, and forgiving, are like goddesses (Bacchetta, 1993; Menon: nd). The *S_dhav_* goes on to emphasize, "character is everything."

The *S_dhav_* asserts that the advertisements you see on television indicate the corruption of the west as opposed to the good character of Hindu women represented in the mythical figures of Sita and Savitri. Savitri sacrificed her own life to protect her husband, a story that not only indicates the Brahmanical values underpinning much Hindu nationalist thought where a woman must serve her husband to find her own fulfilment, but also suggests her bravery and her goodness, in that she was willing to sacrifice herself for the protection of others. Sita, the wife of the God King Ram in the great Hindu epic the *R_m_yan_*, is another brave and virtuous woman. She too, in many versions of the epic, repre-

sents the ideal woman of Brahmanical Hinduism, suffering all the hardships of
exile in the forest because the proper place of a woman is by her husband's side.
The story conveys some important lessons about ideal Hindu womanhood without actually preaching to the participants, but rather, by simply telling them a
story.

The songs too were important vehicles for the dissemination of ideology
addressing as they did many of the themes critical to Hindu nationalist discourse, particularly of the role women are to play in the larger movement. The
fact that I can still remember off the top of my head the tune and a few of the
lines of the songs one and a half years later, but need to refer to my field notes
for the content of the lectures, stands testament to the efficacy of the songs, and
the efforts taken to make sure we memorized them at the _ivir. I was, even next
to Bina, the worst at memorising the songs. We would be asked to stand up and
sing the song one by one to make sure we had learnt it. When my turn came I
had that uncomfortable feeling of being the class dummy. Below I provide a
rough translation (rather than a poetic one) of the second verse and the chorus of
one of the songs we were made to write out and then memorize in the song and
prayer learning class every morning:

> Chorus: Hindus awake! Nation awake!
>   Self-respect and determination awake
>   Hindus awake! Nation awake!
>   Self-respect and determination awake
>   The true *dharm_'s*[8] victory will be immanent
> Verse 2: Love will awake between friends
>   The stream of one blood will awake
>   It is one nation and one culture
>   It is one nation and one culture
>   Our duty to ourselves must awake[9]

This song clearly embodies the ideological principles of Hindu nationalism
containing these in a catchy, if repetitive, tune. It powerfully links the nation
with religion, morality, blood and culture. This is a nation marked, as the next
verse claims, by the awakening of the God Ram (legendary for his sense of
duty) in every one of its members. It powerfully conveys that it is the duty of an
already existent nation of Hindus, who are related by blood, morality, love and
culture, to wake up from a state of passivity to one of self-respect and self-
determination. The song does not dwell on tensions between Muslims and Hindus, or the dangers of Christian proselytising. However, it still manages to convey the righteousness and morality of the Hindu nation and its natural claim to
the territory of India. Its power rests in its ability to legitimise the claims of the
Hindu nation without recourse to vilification of an "other".

Yet, we must not assume that the audiences, "unproblematically accept the
subject positions created by dominant ideologies" (Mankekar, 1999, p. 255).
Stuart Hall asserts; "there is no necessary correspondence between encoding and
decoding" (1980, p. 135) and "any already constituted sign is potentially trans-

formable into more than one connotative configuration" (1980, p. 134). As Purnima Mankekar argues, responses of viewers to texts such as television, or in my case folklore (songs and stories), "are refracted by the discursive contexts in which they live" (1999, p. 253) and are thus always "negotiated readings" (1999, p. 254) shaped by local conditions.

The idea that Hindus need to reject "foreign" cultures and embrace their own heritage was critical throughout the camp, but was nowhere more apparent than in the games chosen by the Samiti for the participants to play. Even games that perhaps have been imported from the west are consciously reformulated so that they appear to be Indian. Many of these games are aimed at increasing the stamina and physical fitness of the participants, suggesting a tendency in nationalist discourses to use "athletic metaphors ... to portray an image of the state as fit, virile, and heroic" (Alter, 1994, p. 558). Samiti members also emphasised that the games taught the participants concentration. These are key to what the Samiti wants its members to learn in terms of its overall goal of making women physically and mentally strong. However, also embedded in these games are lessons about competition and aggression that speak to the larger Hindu nationalist agenda of making Hindus able to defend their nation and culture. Additionally, games like "*Jai Shivaji*" (described below) by indexing key figures in Hindu nationalist versions of history also have the effect of referencing their constructions of the past without explicitly talking about them. It is important to contrast the spirit of competition and general disorder of these games with the attention given to order and discipline during the various exercise drills that Samiti members are also required to participate in. The importance of discipline, particularly disciplined bodies, has been noted by Joseph Alter in his study of the all-male RSS training camps (1994). Apart from the game sessions, the Samiti training camps too are marked by the efforts taken by the leaders to impose order, obedience and discipline. Often this manifested in the attention given to making sure that people stood or sat in perfectly straight columns and rows during exercise drills (see Alter 1994), during song, story, prayer, and lecture sessions, as well as in the efforts taken to ensure order and tidiness in the rooms where even shoes had to be arranged in a perfect line neatly against the wall.

Below I describe a few of the games played through the course of the camp. With the exception of "*Jai Shivaji*" which was played during the morning __kh_, the games were mostly played during the evening __kh_s. The focus of the morning __kh_ was on endurance and fitness training as well as learning military commands, while the focus of the evening __kh_ was on playing games.

*Murg_* (chicken): In this game the participants were divided into two teams. Each had to stand in rows facing each other with a gap of about six feet in between the rows. Each person was given a number that reflected her position in the row (see Figure 1). Thus number 1 in team A was the first person in the row and team B had a person across the gap with the same number. Payal would call out a number chosen randomly and then the two people (one from each team) with that number had to come into the gap hopping on one foot with both hands behind their backs holding up the other foot. The idea was to see which of the *murg_s* (chickens) could knock the other one off balance so that she was forced to put the other foot down. This game is similar to "Steal the Bacon."

Figure 1
Formation of the Game of *Murg_ (Chicken)*

Team A:	1	2	3	4	5	6
	X	X	X	X	X	X
			GAP			
Team B:	X	X	X	X	X	X
	1	2	3	4	5	6

Tanks[10]: here everyone got into groups of five and formed a tight circle holding on to each other with arms around each other's waist. There were about five or six groups. When Payal blew on her whistle all the groups (each, a tank) hurled themselves at the others with the objective of making the other groups fall or break up. The team/tank left intact and standing at the end was the winner. This game could become very rough. Maya hurt herself quite badly in this game. Her big toe nail came off and her toe was bandaged for the rest of the _ivir. However she still continued to participate in some of the more peaceful games and in the morning __kh_s the next day.

    Kis_n aur lomd___ (the farmer and the fox)[11]: In this game all the participants were made to stand in a large circle. One person was made the *kis_n* (farmer) and the other was the *lomd___* (fox). The object was for the farmer to catch the fox. If the fox was caught then she became the farmer and had to now chase the new fox (the former farmer). At any point in the chase the fox could tap on the shoulder of (usually ended up being more like a shove) anyone in the circle and make her the fox instead. The farmer would have to keep running until she caught the fox. If she could not catch the fox then Payal would ask the fox to allow herself to be caught so that the farmer could rest. The chase was around the circle, and often entailed weaving in between the girls standing in the circle.

    *Tot_ aur pinjr_* (the parrot and the cage): In this game the participants had to get into groups of three. Each group of three had to stand in their group but also structurally in a larger circle (See Figure 2). Two of them formed the cage by holding hands and the third – the parrot – stood in the middle. One person was made the catcher and another was made an un-caged parrot. The catcher ran around and through the circle trying to catch the parrot. The parrot could either run or enter into any of the cages to escape. At this point the parrot that was already in the cage would have to leave it and start running. If the parrot was caught then she became the catcher and the catcher became the parrot and could escape into one of the cages. This was a fast paced game and, like the previous game, entailed a good deal of fast running and a great deal of quick turning. One girl, who was unfortunately not very fast or dextrous, ended up being the catcher for so long that she fainted from exhaustion.

Figure 2
Formation for the Game of *Tot_ aur Pinjr_* (The Parrot and the Cage)

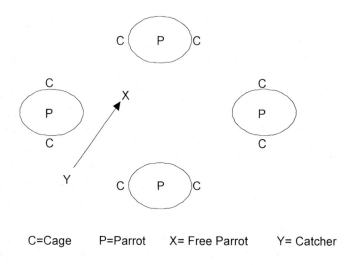

C=Cage   P=Parrot   X= Free Parrot   Y= Catcher

Number making game: In this game all the participants were made to run around in a circle. When Payal called a number everyone had to quickly stop running and join up in groups of that number. Those who couldn't were out.[12]

Jai Shivaji (Victory to Shivaji): In this game the participants had to run around in a circle. Payal stood in the centre of the circle and would call out "*jai Shivaji*" and we would have to answer "*jai Bhavani*". Meanwhile Payal called out the pace of our running by calling out in Hindi "one two, one two". The pace of our running depended on the speed at which she said this. If she said "*jai Bhavani*" instead of "*jai Shivaji*" then we would have to respond with "*jai Shivaji*" and run in the opposite direction. Likewise if she now said "*jai Shivaji*" we would have to change direction.

While this game was geared to improving endurance and to teaching us to pay attention to commands (the game ended a two hour __kh_ focussed on learning various military commands), the choice of terms used is interesting. Shivaji was a seventeenth century Hindu king much revered by the Hindu nationalists as being responsible for founding the Hindu nation despite the strong presence of the Mughal (Muslim) empire. Bhavani was the Hindu goddess that was his family deity and allegedly a motivating force for Shivaji. This game alludes to and therefore reinforces these characters, whose accomplishments and specifically Hindu nationalist constructions had already been introduced to the participants at the camp, without resorting to a lecture on who they were.

In addition to building up strength and stamina and urging people to be competitive and aggressive, many of the games are also importantly about concentration. In a __*kh*_ that I had attended in September with Payal, she had explained another game she had taught them by asserting: "concentration and the ability to pay attention is important in all our lives and in almost every task we do. So it is important that we all learn this" [17.9.99]. The games are a means to teach this lesson. This is particularly significant given that the Samiti feels that women have become physically and mentally weak and the reason for the __*kh_s* is to create an organization of women who are mentally and physically strong and able to defend their nation, values, and culture. Additionally, although many of the games may have their counterparts in the West, it is no coincidence that they appear on the __*kh*_ program in their indigenous guise. It is not simply chance that the girls are not playing hopscotch or some other popular western game. Part of the Samiti's agenda is to convey that these games are Indian games (although this may not be the case) and India has no need for Western imports. As Joseph Alter has asserted, a great deal of the physical regimen in the __*kh*_ format, including the physical exercises and the attention to discipline, have been imported from the West and yet are presented as Indian (Alter, 1994, p. 567).

The other games are also about concentration, endurance, and fitness but interestingly, they are also very much about competition, and, for the most part, are fairly aggressive. The game "tanks" is particularly indicative of this aggression and the participants were very enthusiastic about charging with as much speed as was possible without breaking the circle and hurling themselves at the other tanks to knock them over or break them up. This would result in people literally being knocked over, stomped upon, and crushed by the aggressing tank. What is interesting about the rough and aggressive nature of these games is the way they contrast with the order and discipline required of participants during the rest of the camp.

Felicia McMahon and Brian Sutton-Smith have argued, "children are members of the most politically powerless group" (1995, p. 299), and this is certainly the case in the strict hierarchical organization of the Samiti training camps. The game sessions become an opportunity for them to break away from the rules that structure their interactions with others in the normal course of the camp and express behaviours and emotions that, although limited by a new set of rules, would not have been acceptable outside this context (see Sutton-Smith 2001a). This is all the more poignant in the case of young middle-class Hindu girls who may not be allowed to engage in such aggressive physical behaviour in the normal course of their lives. Games become a space when they can legitimately act in these ways. Games, like those organized in the Samiti training camps, become a space where they can "wield power to affect others," "role play," "belong to a positive reference group," receive "rewards and encouragement for such play," "feel in control of aggression" and so on (see Goldstein 1995, Table 6.4). As

Sutton-Smith has argued play provides a context within which children can struggle "against their ... powerlessness with displays of fictional fortitude" (1995, p. 285). Speaking of children's folklore McMahon and Sutton Smith assert that it is critical to see in children's folklore "the power the children exercise over each other, and the power they seek in their relationship to adults, mythical or real" (1995, p. 299). They go on to argue that play not only "models and manifests power relationships" but also critically "expresses power and power-lessness by being subversive of the normative culture of which it is a part" (McMahon & Sutton-Smith, 1995, p. 301).

While much that is being conveyed by the games echoes themes that have been covered in the lectures, in this new form the message is much more subtle. Participants like Divya and Bina who kept a sullen distance from all the other girls during the various lecture sessions, and girls like Gauri who challenged them outright, did not question the games themselves and in fact participated in them enthusiastically. Bina and Divya, who had consciously stuck together throughout and maintained their distance from the other participants, were enthusiastically a part of the group during these games. It was the only time that they really interacted with the other members. Bina in fact asked Payal after the __*kh*_ was over whether they could keep playing games, but Payal said that we had to stick to the program. The games, more than any other activity of the camp, created the communitas (Turner, 1969) that permitted the participants to transcend the various ideological, social and individual differences between them in the excitement of the chase or the unity of team competition. Yet this sense of community did not necessarily include the adult leaders of the Samiti who were organizing the camp but, more often, just the participants who were mostly the children along with six adults (Divya, Bina, Anshu, Ruchika, Rajani and myself). While Payal also played some of the games with us, the girls were particularly competitive with her, although they never managed to get her out or catch her as the case demanded. While the Samiti's goal was to create a community organized by hierarchical relationships, the participants challenged the existence of this community in play by trying to defeat Payal. Apart from the enthusiastic manner in which she was chased, the participants would also gleefully cheer the catcher on (much more than in other cases) when she happened to be chasing Payal.

On the last day of the camp when I was talking to Divya and Bina, Bina told me that she had only come for the _*ivir* because Divya was coming. She added that she disliked all the things that were being said about Muslims and Christians but she really enjoyed the games. She wished that they only had games and not the *baudhik* sessions. Interestingly, despite her dislike of the ideology of the movement, Bina agreed to attend a festival that Payal was organising in mid January, *makar sankr_nt_*,[13] in which good food and lots of fun games were promised. Thus the games become an important means through which new members are recruited. This is no where more apparent than in the words of Sneha, a nineteen year old girl who is currently with Durga Vahini, one of the other women's wings of the movement. She said that she had attended her first _ivir_ when she was ten years old and she began working with Durga Vahini when she was fifteen. She first went to the _ivir_ with her aunt (the *S_dhav_*, female ascetic, mentioned above) and then got so interested in what they were doing that she decided to commit to working for them. She said; "first it was all fun and games. Then when I was older I started listening more carefully to the *baudhik* lectures and understood the importance of what was being said" [14.10.99].

CONCLUSION

It is precisely the ability to disseminate the values, ideology and morality of Hindu nationalism without raising anyone's hackles that makes play a powerful tool in the hands of Hindu nationalist women at these training camps. While participants voiced their protest at the lectures, none of them expressed reservations about the play sessions either publicly or to me. The stories and songs were important vehicles for conveying nationalist ideology. For instance, the story about Akbar purports to be just a tale about a historical figure; however, it manages to reinforce Hindu nationalist constructions of Muslim rule throughout Indian history, without explicitly talking about Muslims and Hindus. The song conflates Indian identity with Hindu identity without explicitly calling Christianity and Islam foreign. It also clearly conveys that it is the patriotic duty of all Hindus to fight for this Indian/Hindu nation.

The games were by far the most seductive recruiting strategy deployed by the movement. While Bina and Gauri were both unhappy with what was being directly said about Muslims and Christians during the various lectures that they were made to sit through, neither of them contested the legitimacy of the games being played and the subtle messages that were coded in them. They both agreed to attend a subsequent event organised by the Samiti because of the seductive power of the games as well as simply because their friends were going to be there. It is the ability of the Samiti to appeal to both sides, those who are in it for fun, and those who are in it because they believe in its ideology, that allows it to ever expand its base and bring in new recruits to its organization. Yet it is also important to recognize that the participants brought their own interpretations to the various forms of play and sometimes subverted the goals of the movement. This was apparent in the way their play during the game sessions fundamentally reversed all the requirements for order underscored by the movement. It was during the games when the participants, albeit unsuccessfully, tried to challenge Payal's authority and power over them by defeating her in play. Yet, one has to be careful of romanticizing these acts since this subversion too occurred within the context of a specific space and opportunity created by the movement for precisely this inversion of order.

NOTES

1. Funding for this research was provided by the American Institute of Indian Studies. The author gratefully acknowledges the comments and suggestions of Susan Wadley, Felicia McMahon, Mark Hauser. Keri Olsen and an anonymous reviewer. All names used in the article are pseudonyms to protect the identity of the individuals. Personal details have also been altered to protect the identity of the individuals. Parts of this article are from a Chapter in my dissertation (Menon: nd).

2. In the longer camps they include martial arts training as well as instruction in fighting with *l_th_s* (bamboo poles) and daggers.

3. While it is true that the *Koran* says that a man may have up to four wives, it also says that a man can only take another wife if he is willing to treat both wives equally. It goes on to say that a man cannot possibly treat both wives equally. However, in practice, Muslim men do indeed have more than one wife in parts of India.

4. A North Indian dress consisting of a long tunic worn over loose pants.

5. *Ant_k_in_* is a well-known game played throughout India and is usually played with using songs from Hindi films. Basically one team sings a few lines from a song. The other team now has to sing a song starting with the last letter of the previous song. A team loses if they are unable to think of a song starting with the letter assigned by the other team. During the camp we had to play *ant_k_in_* using patriotic songs rather than Hindi film songs and I was completely out of my league.

6. Hindu nationalist groups in India are so against the observance of Valentine's Day, that in February 2001, the Shiv Sena, one of these groups, headed a raid into greeting card shops in Mumbai, and destroyed all the Valentine's Day cards.

7. *Bh_rat M_t_* has been accorded the status of a goddess in Hindu nationalist discourse (see Bachetta, 1994).

8. *Dharm_* can be used synonymously with religion but can also be used to refer to the moral order and cosmic duty.

9. *Hindu Jage, Desh Jage, Sw_bhim_n Sankalp jage (2)/ Satya dharma k_ vijay sunishchit/ Bandh_ Bandh_ me py_r jage/ Ek lah_ k_ dh_r jage/ Ek desh hain, ek sanskrit_ (2)/ sva kartavya vich_r jage.*

10. The game was referred to by the English word "Tanks."

11. Two readers have suggested that this game is similar to Duck Duck Goose played in the United States.

12. As one reader has suggested, this game is similar to Musical Chairs.

13. *Makar sankr_nt_* is a harvest festival celebrated in many parts of North India. The day marks the beginning of the Sun's movement north from the Tropic of Cancer to the Tropic of Capricorn. Harvest festivals are celebrated at the same time in other parts of India such as *pongal* in Tamil Nadu and Andhra Pradesh and *lohr_* in Punjab.

REFERENCES

Alter, J. (1994). Somatic nationalism: Indian wrestling and militant Hinduism. *Modern Asian Studies, 28*(3):557-588.

Appadurai, A. (1997). *Modernity at large: Cultural dimensions of globalization.* New Delhi: Oxford.

Bacchetta, P. (1993). All our goddesses are armed: Religion, resistance and revenge in the life of a militant Hindu nationalist woman. *The Bulletin of Concerned Asian Scholars, 25*:(4): 38-51.

Bauman, R. (1977). *Verbal art as performance.* Rowley, MA: Newbury House Publishers.

Brow, J. (1996). *Demons and development: The struggle for community in a Sri Lankan village.* Tucson, AZ: University of Arizona Press.

Gold, A. G. & Gujar, B. R. (2002). *In the time of trees and sorrows: Nature, power and memory in Rajasthan.* Durham, NC: Duke University Press.

Goldstein, J. (1995). Aggressive toy play. In Anthony Pellegrini (Ed.), *The future of play theory: A multidisciplinary inquiry into the contributions of Brian Sutton-Smith.* (pp. 127-147). Albany, NY: State University of New York Press.

Hall, S. (1980). Encoding/decoding. In S. Hall, D. Hobson, A. Lowe & P. Willis (Eds.), *Culture, media, language: Working papers in cultural studies, 1972-79* (pp. 128-138). London: Hutchinson.

Hancock, M. E. (1999). *Womanhood in the making: Domestic ritual and public culture in urban south India.* Boulder, CO: Westview Press.

Loy, J. W. & Hesketh, G. L. (1995). Competitive play on the plains: An analysis of games and warfare among Native American warrior societies, 1800-1850. In Anthony Pellegrini (Ed.), *The future of play theory: A multidisciplinary inquiry into the contributions of Brian Sutton-Smith.* (pp. 87-99). Albany, NY: State University of New York Press.

Malkki, L. H. (1995). *Purity and exile: Violence, memory, and national cosmology among Hutu refugees in Tanzania.* Chicago: University of Chicago Press.

Mankekar, P. (1999). *Screening culture, viewing politics: An ethnography of television, womanhood and nation in postcolonial India.* Durham, NC: Duke University Press.

McMahon, F. R. and Sutton-Smith, B. (1995). The past in the present: Theoretical directions for children's folklore. In B. Sutton-Smith, J. Mechling, T. Johnson & F. McMahon (Eds.), *Children's folklore: A source book* (pp. 293-309). New York: Garland Publishing.

Menon, K. D. (nd). *Dissonant subjects: Women in the Hindu nationalist movement in India.* Unpublished doctoral dissertation. Syracuse University.

Narayan, K. (1989). *Storytellers, saints and scoundrels: Folk narrative in Hindu religious teaching.* Philadelphia, PA: University of Pennsylvania Press.

Schwartzman, H. B. (1995). Representing children's play: Anthropologists at work. In Anthony Pellegrini (Ed.), *The future of play theory: A multidisci-*

plinary inquiry into the contributions of Brian Sutton-Smith*. Albany: State University of New York Press.

Singer, M. (1972). *When a great tradition modernizes: An anthropological approach to Indian civilization.* London: Pall Mall Press.

Sutton-Smith, B. (1995). Conclusion: The persuasive rhetorics of play. In Anthony Pellegrini (Ed.), *The future of play theory: A multidisciplinary inquiry into the contributions of Brian Sutton-Smith.* Albany, NY: State University of New York Press.

Sutton-Smith, B. (1997). *The ambiguity of play.* Cambridge, MA: Harvard University Press.

Sutton-Smith, B. (2001a). Emotional breaches in play and narrative. In A. Goncu & E. L. Klein (Eds.), *Children in play, story, and school.* NY: Guildford Publications.

Sutton-Smith, B. (2001b). Reframing the variability of players and play. In Stuart Reifel (Ed.), *Theory in Context and Out. Play and Culture Studies, Volume 3* (pp. 27-49). Westport, CT: Ablex Publishing.

Turner, V. (1974). Liminal to liminoid, in play, flow and ritual: An essay in comparative symbology. *Rice University Studies, 60.* 53-92.

Turner, V. (1975). Ritual as communication and potency: An Ndembu case study. In Carole Hill (Ed.), *Symbols and society: Essays on belief systems in action.* GA: University of Georgia Press.

Turner, V. (1969). *The ritual process: Structure and anti-structure.* Chicago: Aldine.

Chapter 7

Ludic Pathologies and Their Links to Play: Cultural and Neurocognitive Perspectives

Donald E. Lytle

Humans sometimes exhibit interesting, quirky behaviors that are often irrational and sometimes pathological and even deadly. While all the behaviors are impulsively compulsive, they tend to manifest themselves differently in various cultures. Eating disorders among women in Western nations (anorexia nervosa and bulimia), spirit possession in women living in the Horn of Africa and Eygpt (*zar*), and Arctic hysteria in Greenland Inuits (brooding followed by extremely violent behavior) are diverse examples of these apparently irrational behaviors. All are idiopathic in origin and onset.

These so-called, culture-bound behaviors, are more accurately described as ludic syndromes[1] because they are all characterized by a group of identifiable symptoms that are irrational, inane and playfully derived. None have easily identifiable cultural, ethological or biomedical mechanisms, and anthropologists, and Western medicine practitioners and researchers have been unsuccessful in identifying a common link between these syndromes and the mechanisms for their onset and symptoms.

Latah is a physical and behavioral response to being startled and is particularly common among older women who live in Malaysia and other Southeast Asian countries. It is characterized by hypersensitivity to sudden fright, and is often accompanied by echokinesis (imitation of other's movements), utterance of sexual obscenities, command obedience, and dissociative or trance-like behavior. The symptoms of Latah are neurocortically derived but are mostly playful and tend to occur during the play of others. Latah may therefore provide insight to the universal function and genesis of play.

CULTURE-BOUND LUDIC SYNDROMES

Culture-bound syndromes, introduced by Pow Meng Yap (1969), are defined as a collection of signs and symptoms of 'disease' restricted to a limited number of cultures relative to certain psychosocial features (Prince, 1985). Hughes (1985) presented a "Glossary of 'Culture-Bound' or Psychiatric Syndromes" which included over 180 distinct cross-cultural displays under different names. The listings for culture-bound syndromes are long and confusing and related to non-existent factors, such as the "vapors" or one or more ancient bodily "humors." Lack of clarity and classification confusion are also seen in the considerable overlap of symptoms within classified culture-bound syndromes. For example, the "anger syndrome" of Korea (*hwa-byung* or *wool-hwa-bung*) involves insomnia, fatigue, panic, fear of impending death, dysphoric affect, indigestion, anorexia, dyspnea, palpitations, generalized aches and pains, and a feeling of a mass in the region above the stomach (Winzler, 1995).

Culturally-specific idioms of distress and "disease" that appear in the culture-bound and folk taxonomy literature add to the confusion. The relationship of phenomena related to soul, mind, and body is difficult to reconcile. For example, the 'evil eye' is a common idiom of disease, misfortune, and social disruption throughout the Mediterranean, Latin American, and Muslim countries. Spells and related customs, common in the Southern United States and Caribbean, are often sensationalized and made magically mysterious and are used to explain illness. The process involves hexing, witchcraft, voodoo that can trigger sudden collapse and fainting. In Ethiopia, Somalia, Egypt, Sudan, Iran, and elsewhere in North Africa and the Middle East, *zar* involves dissociative episodes often related to possessing spirits. This has evolved into a dance among young adults who go to disco-type clubs to dance *zar* using colored scarves representative of the demons who possess the women.

There are many cross-cultural behaviors associated with rageful emotions. The most notorious is *amok* or *mata elap* associated with Malaysia, and so-called Arctic hysteria or *pibloktoq* of Greenland natives. Like the dissociative extreme excitement of the Nicaraguan Miskito Indians (*grisi siknis*), there is behavioral similarity in *cafard* or *cathard* (Polynesia), *mal de pelea* (Puerto Rico), *iich'aa* (Navaho), *boufée deliriante* (West Africa and Haiti), and *ataque de nervios* principally reported among Latinos from the Caribbean, and also found among many Latin American and Latin Mediterranean groups.

There is hypochondriacal unease with semen-loss syndrome in India (*dhat*), Sri Lanka (*sukra prameha*) and China *(shenkui)*. An episode of sudden and intense anxiety that the penis will recede into the body and possibly cause death occurs throughout south and east Asia and is named *suo yang* in China, *jinjinia bemar* in Assam, *rok-joo* in Thailand and *koro* throughout Malaysia. Bilis and colera, both described in terms of hot and cold, are common Latin American idioms of distress as well as being cultural explanations for physical and/or mental illness because of extreme emotional upset. For this same group, *susto*, is an illness attributed to a frightening event that causes the soul to leave the body.

This chapter will demonstrate how the imprecision, elusiveness and ambiguity of play may be associated with all of these vagaries. Latah, (hereafter capitalized to refer to individuals and behavior), will be analyzed as a ludic syndrome culturally informed, and uniquely playfully constructed given its emotional neurobiological foundation. Play's evolutionary neurocognitive foundation and function will explicate how these ludic syndromes are manifest.

Latah

To conceptualize two common views of Latah one must first know how it manifests itself in human behavior for scientists have long struggled with a satisfactory explanation. Latah is an illustrative and interesting case that affects individuals in Southeast Asia among the Malay and the adjacent and immigrant peoples in Borneo and Java. Those with Latah may be socially marginalized men, but far more typically are highly esteemed well-loved middle-aged and elderly women in the community who are proud to be recognized as having and being Latah. It is instigated in association with others by a startle reaction that also frequently involves blurting out offensive words or phrases usually sexual in nature. It is not uncommon that the words 'penis' or 'vagina' would be called out. There's also imitative behavior often of other individuals present, and in addition commands may be given by others to coerce obedience of the subject (Simons, 1983, 1996). It has also been construed as a form of trance, and after the event most Latahs claim to have no memory of what they said or did.

From a cultural perspective there is disagreement and theoretical speculation as to the onset of Latah. Some like Murphy (1976) attribute the first appearance of Latah to the European intrusion in Malay life and make a case for it as a native reaction to rapid social change. Others, such as Kenny (1978, 1990) however, show that Latah has been an important part of Malayan culture for a time before the nineteenth century. Adding somewhat more confusion to the mix, Margaret Mead and Reo Fortune in their investigation of the Tchambuli people of Papua New Guinea during the 1930s reveal the underpinnings of play within the manifestation of Latah (Mead, 1935).

Those preferring a cultural explanation posit "that Western psychiatry is increasingly being forced to deal with the role of culture in creating mental illness" (Osborne, 2001, p. 100). Canadian anthropologist Michael Kenny even goes further with the sociocultural-psychological view that Latah is related to ritual and "local witchcraft beliefs, to midwifery, to shamanism, to folk art, and to fundamental ideas pertaining to the gaining of religious insight and power through loss of self" (Kenny, 1983, p. 160).

Similarly confusion exists among theorists as to classification and geographic specificity, for startle effects and similar characteristics of Latah appear throughout the world. Even the behavior patterns of obscenities and imitations from individuals with Tourette's Syndrome are strikingly similar to those of Latah: So much so that Georges Gilles de la Tourette explicitly compared this famous syndrome to Latah (Tourette, 1884). More of the comparison will be discussed later in this chapter. Winzeler (1995, p. 131) documents what he calls "true Latah," with manifested sexual obscenities, loss of consciousness, imitative and command-obedience behaviors occurring in Thailand (*bah-tschi*) and the Philippines (*mali-mali*) and across far northern Asia from Japan (*ainu* or *imu*) through Siberia (*myriachit*) to Scandinavia (Lapp panic), and among French-Canadians in Maine (jumping). In 1832, Charles Darwin on the Beagle voyage encountered imitative behavior from the Tierra del Fuego Indians. With no mention of startle, his prejudicial journal entry generalized that "all savages appear to be excellent mimics" (quoted in Winzeler, 1995, p. 38), which leaves ambiguous the possibility that he was also witnessing Latah.

So the relatively vast discussion surrounding Latah has "been animated especially by reference to a seeming paradox" (Winzeler, 1995, p. 130). How can it be culturally derived, yet occur in diverse cultures throughout the world? In an often referred to 1968 paper, Hildred Geertz, capsulated the problem. After extensive study in Java, she argued that the Javanese form of Latah reflected regional values and customs centered upon a colonial hierarchical social order, yet she also knew that similar behaviors occurred in distant geographical settings quite apart from Southeast Asia.

From a contrasting biological viewpoint, researchers have focused on the "almost invariable" and "innate reflex" startle response which universality underlies the Latah syndrome (Landis & Hunt, 1939, p. 146). In similar Westernized fashion after the middle of the nineteenth century when European observers reported on the Latah-like behavior, it "made its way into the second- and third-hand literature of manuals of tropical medicine and psychiatric handbooks as a recognizable (if apparently incurable) mental disease of uncertain nature" and that's where it tended to remain (Winzeler, 1995, p. 3). Thus Latah became a standard textbook case in psychology, anthropology, medical anthropology, culture and personality studies, where it came to be known as an "exotic psychosis or neurosis" (Winzeler, 1995, p. 3). This was in spite of the fact that when the supposed patients were given general, cardiac, and neurologic examinations, complete blood counts, serum chemistry, magnetic resonance imaging scans of the brain, electroencephalograms, electrocardiograms, and echocardiograms, they were found to be otherwise normal. None the less the term "disease" and the naming of this type of behavior as "reaction" or "syndrome" is still commonly used to describe "patients" who are "diagnosed" and supposedly "suffer" with this disorder.

Psychiatrist Ronald Simons presented a compromise position between the cultural specific emphasis and the biomedical model for explaining Latah. First presented in a 1980 paper, *"The Resolution of the Latah Paradox,"* he explained the phenomenon as a culture-specific exploitation of a neurophysiological potential. This was like the bridge between what Clifford Geertz discussed as the biological 'blink' and the symbolic and cultural 'wink.' However, relative to the function of play and its presentation, the biological reaction versus the cultural relation views of Latah are off the mark. Rather than focus upon the Western biomedical disease or culture-bound syndrome approach, and the contrasting culture-free or psychosocial, anthropological explanation, perhaps an alternative explication for Latah and similar behaviors seen within cultures is an interpersonal, intrapersonal, and neurologically grounded play-based one.

The Playful Latah

Kenny (1983) hints at the important role of play when he explains Latah in light of the "presence of liminal marginality," the "social nature of the display," and as the symbolic expression of loss of self through "disordered behavior." He writes of "a human repertoire of concepts pertaining to order disorder, and self-identity" (Kenny, 1983, p.161). Further he prefers to replace the culturally variant term latah with "performance" and writes of "involuntary mimesis" and obscenity of the Latah which, virtually by definition, stands in a mirror relationship to specific norms of order, decency, courtesy and poise. Hahn (1995) believes this view has gone too far. Critiquing Kenny's dramaturgical notion he writes that Kenny "replaces biomedical reductionism with theatrical reductionism" (Hahn, 1995, p. 46). However, maybe Kenny didn't go far enough in characterizing a ludic syndrome. Performance studies scholar Richard Schechner cautioned play researchers to "investigate playing, the ongoing underlying process of off-balancing, loosening, bending, twisting, reconfiguring, and transforming - the permeating, eruptive/disruptive energy and mood below, behind, and to the sides of focused attention" (1988, p. 18). He could have been describing Latah.

Since Latah is always brought on by interaction with others, the foundation is ripe for play, and thus upholds Bateson's notion that play must involve "the establishment and exploration of relationship" (1980, p. 151). Unlike the often-marginalized 'victims' of hyperstartle syndrome in the West, Latahs in Southeast Asian seldom seek treatment, are very accepted, and even celebrated for their behavior. They are fun to play with. Friends and family members, especially grandchildren of Latahs deliberately provoke a startle response. Some common methods involve pokes to the ribs or tin pots thrown behind the Latah's backs. A grandson of a Malay Latah reports, "I just clap my hands behind her back, and she starts throwing stuff around the room" (Osborne, 2001, p. 100). Latahs are easy targets, for over time they become so sensitive that trances can be triggered by something as mundane as a falling coconut. As entertaining burlesque akin to a clown or magician, many have been invited to birthday celebrations.

As a complex form of social play Latah is fun and funny. In Thailand it is commonly referred to as "tickle madness" (*bahtschi*). Bartholomew (1994), who married into a Malaysian extended family with many Latah cases, presents a case that it is best regarded as an attention-seeking measure. For the relatively staid Malay, it is grand innovation for onlookers to hear shocking obscenities and see mocking behaviors. An entertaining 69-year-old Latah woman usually dances wildly for up to 30 minutes. A Latah in Borneo said that she imitates "everything I see around me: Even the TV." When asked about the danger of that, she titters and says, "I try not to watch too much wrestling" (Osborne, 2001, p. 100).

Humorous teasing, joking and shocking, bawdy insults are Latah's ritualistic hallmark, and therein preserve and display the marked playful elements of creativity and the unexpected (Bateson, 1980). Further, the dramatic action involves inversion and parody, often going against normative mores and behavioral standards. After all, the initial retorts after the startle are usually sexual taboo topics. These expressions are certainly not appropriate for young women and even for men in the Malay culture: However, older women have a bit more freedom in this regard. "Latah behavior is partly a matter of contradiction but it is also partly a matter of exaggeration" (Winzeler, 1995, p. 130): Thus, the playful quality of hyperbole is consistently exhibited.

Interspecies play may be common relative to Latah. A report of a middle-aged Latah woman in a Malaysian jungle hamlet involved her relationship to cats. Her husband explains that "It's cats that get her the most. They make her more latah than anything" (Osborne, 2001, p. 100). Victor Turner (1967) writing of the Ndembu shows that playing with opposition to the familiar human form is common in cultural rituals, and in stories associated with Trickster.

Trickster and Latahs

Tribal tricksters are part of the human story throughout the world and have been for a very long time. Notable for North Americans are the clever tricksters of Native American origin embodied by Coyote, Raven or Iktomi. In Norse mythology, it's Loki, in Africa, he's called Anansi/Ananse, Eshu, Legba or Spider, and in Asia and South America, she's known as the Fox. Unlike Latah individuals, the trickster is usually in animal form, desirous of food and can shift shapes and gender easily. What is common to the Latah phenomenon is metaphorically this disguise element, "at once be [ing] himself and not himself" (Hyde, 1998, p. 54), and more tangibly, the relationship to sexual behavior mischievous prankishness and the surprise that surrounds them. Both are scandalous disrupters of the social order. The improvisational and dance-like nature of both is another similarity. For the Latahs the natural improvisation becomes performance art and the 'dance' per se is reciprocally related to the behavior of others. For the Tricksters like Ananse there is often music playing and the correspondent natural rhythmic movement may even involve tumbling and cartwheels.

Whereas the trickster stories involve cosmological truths, creation, death and rebirth they are similar to the Latahs in that they also have moral endpoints and are pastime tales designed for entertainment and instruction. Metaphorical in the former and actual in the latter case, those confronted by the Latahs are shown their decided way of being in the world. The same is true of the Brer

Rabbit stories, which emanated from the African-American tales of the hare, Leuk, who provided a means of coming to terms with the bondage predicament of plantation culture (Drame, 1996). Through the playful, the hidden truth of the results of societal practices and policies is revealed. In this way the Latahs as "prophets with a difference" (Hyde, 1998, p. 294) can "speak to the people through the medium of imagination" (Hyde, 1998, p. 295). Epitomized in some poets, ancient tricksters, shamans, court jesters, fools, ritual clowns and individuals with Latah, where did this 'crazy wisdom' and playful action come from?

NEUROCOGNITION, LATAH AND PLAY

As seen in the preceding discussion, in a commonly demeaning fashion the concept of Latah continues to be misunderstood, misappropriated, and misinterpreted. And amongst the discernible playful qualities of Latah, the important underlying causes and function of play is hidden. From a psychotherapeutic perspective the epidemiologic foundation of the playful responses of Latahs has been explained as the screen memory representations and fantasies of what individuals feared or wished might have happened in their past (Stafford-Clark, 1971). This relates to the collective social forgetting practices of Melanesian women in particular, "as a social mechanism of alienation" (p. 430). However, more concretely, and poetically stated as the "embodied trope of the fabricated memory" (Battaglia, 1993, p. 433), Latahs' behavior can be explained neurocognitively. For example, for a long time within traditional transcultural psychiatry beliefs of psychosocial differences prevailed, now biological explanations predominate (Prince, 2000). The mind/body process of those Latah individuals retains the playful and significant emotional repertoires of other humans, however there are obvious and distinct differences in how it is manifest. To see how this may happen and with the caveat that the precise neural architecture and various brain functions have yet to be identified, the sensory, imaginative, emotional and neurocognitive features of the human will be presented next. Also the similar abnormality of Tourette's Syndrome and later, the dissimilarity of autism will be discussed relative to Latah.

The Somatosensory System and Tourette's Syndrome

When humans encounter sensory stimuli and become aware of changes in our own body, without the informational component and possibility of intentional acts, sensation may be considered the most primitive form of mental capacity. Memory, and thought are the imaginative ways in which individuals combine the sensory input with the neurocognitive for the intentional part of the formula. Ben-Ze'ev (2001) explains it this way: "The ability to imagine situations that are different from those presented to the senses is an essential feature of human consciousness in general and of emotions in particular" (p. 191). In this playful process we refer to what is not actually present to the senses. So that in the case of Latah and in the human style of metaphoric thinking, anger involves imagining what should have been, but was not. Mark Twain said that "You cannot depend on your eyes when your imagination is out of focus," and in the startle of the Latah, they "come to their senses" in a sense.

The somatosensory system is complex, but the concept of fixed action patterns (FAPs) will help explicate foundationally what may be happening to the Latah neurocognitively. Remember those many times you have been driving and you don't remember steering and controlling the vehicle over long distances? This was largely due to the reflective and constrained nature of FAPs. Other fixed action patterns are numerous within humans from swallowing to the escape response. Intriguingly they relate to creativity, for learning consists in large part to the adjustment of FAPs, language-use is a complex FAP, and emotions are premotor FAPs (Llinás, 2001). They are generated by the basal ganglia, a large set of subcortical nuclei intimately related to the brain's motor systems. Hence walking, without 'thinking' about the action of arms, legs and feet, is a prototypical example. The capacity of this storehouse of motor programs and the vast overcompleteness of the motor system, allows for completion of a given movement in an almost infinite number of ways

As pointed out earlier there is a compelling similarity between Latah and Tourette's syndrome, and in the latter case there is an excess of very particular types of fixed action patterns (Saba et al., 1998). By contrast and in reverse fashion, with the selective degeneration of a portion of the substantia nigra, one of the basal ganglia nuclei, Parkinson's syndrome is manifest (Olanow & Tatton, 1999). The FAPs of Tourette's patients are seen in the continuous drumming of their fingers, constant arm movement, and/or perhaps unremitting talking. These individuals like those with Latah are witty, probably intellectual, often athletic and respond very quickly to sensory stimuli that relate to moticity, like eye-hand coordination. Llinás (2001) says under these conditions they bring "into focus the issue of automatic motor activity, . . . for the most part it is normally suppressed, and then subject to abnormal, involuntary liberation under very selective pathological conditions" (p. 142).

Individuals with Tourette's syndrome when being stopped or terminating a motor act, are compelled by their neuropathology to continue the act, and often do so through the generation of short, loud words (Llinás, 2001). In the similar case of "ballism" the person completes the action with spontaneous arm flailing (Berardelli, 1995). The Latah individual with a different composition of modular motor acts, incorporates the expletives into the creative, imitative repertoire. This is interesting for normally a human fixed action pattern when interrupted instigates a cortical connection to the large association areas of the cerebral cortex. What is happening neuronally is that the FAPs, implemented at the basal ganglia, are put into context by the reentry of basal ganglia output into the ever-cycling thalamocortical system. A most complex organizational system, the basal ganglia sends and receives information from the thalamus. In fact the intralaminar complex of the thalamus abundantly and directly projects onto the basal ganglia. Therefore there is an active physiological interplay between the self and FAPs. In the case of Latah, biological startle becomes a trigger for a whole host of essentially playful encounters and motoric acts from coprolalia to echolalia. The startle in this way is converted to a prosocial theatric performance rather than an anti-social action as seen with the common episodes of rage attacks in children with Tourettes (Budmana et al., 2003). The primary emotions of the 'old brain' of the thalamus, hypothalamus and especially the amygala all interact to create an existential confrontation with, and manifestation of what it means to be human. What makes this possible neurologically will be discussed next beginning with an amazing discovery.

Mirror Neurons, Imitation, Autism and Latah

Several years ago scientists were studying brain neurons in monkeys for the actions of grasping, holding and tearing, and a most amazing discovery was made by accident. They found that neurons discharged not only when the monkey grasped or manipulated objects, but also when the monkey observed the experimenter or another monkey making a similar gesture (Rizzolatti & Arbib, 1998). The neurons fired when the experimenter, near or far away, made some action (typically a hand movement). Many of the neurons responded selectively, as when the monkey observed one type of action, such as hand grasping or tearing compared to mouth grasping. Some 'mirror neurons' have been found that are highly specific, coding not only the action aim, but also how that action is executed. They fire, for example, during observation of grasping movements, but only when the object is grasped with the index finger and the thumb. Further it has been found that neurons in the ventral premotor cortex of macaque monkeys (*Macaca nemestrina*) discharge not only when the animal performs a specific action or observes the same action, but also when it hears the related sound (Kohler et al., 2002). These audiovisual mirror neurons code actions independently of whether the actions are performed, heard, or seen.

Using noninvasive positron emission tomography scans, these mirror neurons subsequently have been found in humans. Located in the premotor cortex just in front of the motor cortex, is the same mechanism for recognizing the actions made by others. Interestingly, the region in which the mirror neurons are found is proportionately the same size in humans and chimpanzees and are located in the monkey homolog of Broca's area, the language center in human brains. Thus this mechanism provides the neural prerequisite for development of inter-individual communication, via gesture, and finally speech. Human speech evolved from this ancient gestural communication system (Rizzolatti & Arbib, 1998), for it's the oro-facial gestures, not the vocalizations, that link to the monkey's 'pre-Broca' area. Analysis of fossil skulls suggest that as the primate brain evolved, it was building a language mechanism even before there was a larynx capable of articulating speech.

The discovery of the mirror-neuron system and its neurologically-distributed process is so significant that noted neuroscientist, Vilayanur Ramachandran believes "that mirror neurons will do for psychology what DNA did for biology: they will provide a unifying framework and help explain a host of mental abilities that have hitherto remained mysterious and inaccessible to experiments" (Ramachandran, 2000, p. 1). The action of the multimodal, mirror neurons, FAPs and the shared code between perceptions and embodied emotion to imitate actions (and understand them) may lend insight into the Latah and playful action.

From numerous studies and the discovery of the special mirror neurons, which allows us to put ourselves in the place of others, it has been confirmed that humans are special empathetic, playful creatures neurocortically driven. Many theories have surfaced to help explain the neurologically-distributed and representational process concerning our empathy, theory of mind and emotional system. There is the direct matching hypothesis (Flanagan & Johansson, 2003), the hyperbolic discounting theory, simulation theory, and various appraisal models of emotion which hold that emotions are generated when the subject consciously or unconsciously evaluates or appraises the event (see Omdahl, 1995). Imagination is a necessary given of this system which interacts with almost all major brain areas. For example simulation theory is based on the idea that people playfully understand what is going through the minds of others by mentally mimicking what the other is thinking, feeling or doing. The process oriented Perception-Action Model (Preston & De Waal, 2002) broadly integrates the primary neurological functions with the evolutionary and adaptive nature of the representational system. With this model there is a strong overlap with Antonio Damasio's Somatic Marker Hypothesis of emotion (Damasio, 1994), and with his views on the neurobiology of emotion and feeling (Damasio, 1999; Damasio et al., 2000) wherein perception activates one's stored representations and these representations are linked to one's associated feeling states. The entire process ending with the re-activated representations, or what Damasio calls "images," function well in most people, but not well in others.

The varied types of empathy disorders and theory of mind deficits reveal the malfunction of the mirror neuron and FAP system. With their study of failed mu wave suppression, Vilayanur Ramachandran (2000b), and his postdoctoral student Eric Altschuler found indirect evidence that individuals with autism have anomalous mirror neuron systems, so they may not be able to look others in the eyes, may treat others violently as objects, and often may not respond to their own names being called. Individuals with autism typically display affect inappropriately (not necessarily 'flat affect'), and an inability to fully tune into and imitate others (Buten, 2004). With autism the phenomenal and natural experience of mirroring ourselves into others is obviously impaired, and with Latah there is a different neurological configuration.

Latah is developmentally opposite the onset of the majority of empathy disorders like autism which are most severe in individuals who have problems from infancy. Likewise, imitation normally emerges early, yet apropos to Latah, experimental evidence also demonstrates "the need for experience to fine-tune the circuits for responding with the object" (Preston & de Waal, p. 5). Imitative actions are normally centrally inhibited prefrontally and peripherally inhibited with the spinal cord blocking the motor neurons that execute the intended action. This is not the case with Latah where imitative action comes by and large spontaneously.

Although very controversial, imitation of those with autism has been an effective means of allowing breakthroughs in behavioral change. Barry and Fran Kaufman (1995) tell the story of how countless hours of imitation lead their son to a full recovery from autism. Psychologist Howard Buten, an early advocate of the imitative style of treatment, has devoted his life to helping those with autism. Acknowledging the power of imitation and its relation to the depth of imagination, he writes that "imagining is a professional skill as necessary for those working in our profession as it is for actors" (Buten, 2004, p. 167). It undoubtedly helped that Buten is also a professional clown and brings other forms of playful interaction to therapy sessions. This imagining, "the clearest form of empathy" (Buten, 2004, p. 159), is what the human mind/body system does naturally. In Buten's view this commonplace ability, hidden from the autistic, involves "imagination, intuition, inventiveness, [and] playfulness" (p. 166).

In autism the foundation of neurocortical play and its link to affective relationships is masked by repetitive actions and such things as echolalia. This repeating of words either immediately or sometime later, is similar to the flow of 'swear' words uttered by the individual with Latah. Being pre-wired to playfully tune into the actions of others, language/speech, or lack thereof, is involved with Latah and autism and also relates to either the predominance of fixed action patterns in the case of Tourettes or autism or neurocognitively expressed with creative movement repertoires in the case of Latah. Further as Rizzolatti and Arbib (1998) suggest, the neural representation of the mirror neuron system is accessible to other neural systems. For example Ackerman (1990) writes, "there is a deep emotional component to pain" (p. 108). As discussed, the sensory motor system is not limited to controlling movements. It is also involved in imagination: reading and understanding the actions of others – and our selves. For instance in the case of autism it is believed that proper development and social-emotion interest must be met with behavioral responsiveness by the caregiver. Ironically, in the case of Latah, the persons affected are the most experienced caregivers. What are the specific brain mechanisms that allow this affective, imaginative, imitative process to emerge?

A region of the primate insula, the dysgranular field, has a key role in affect generation and empathic resonance, for it connects the limbic areas with the inferior frontal, anterior part of the posterior parietal and superior temporal cortex. This is necessary circuitry for imitation and action understanding as the superior temporal cortex provides an early visual description of the action to be imitated to specific parietal areas with neurons that have the unique function of matching observation and execution of action. Parietal neurons with observation/execution matching properties add additional somatosensory information to the movement to be imitated. Normally developed individuals then code the goal of the imitated action within the inferior frontal cortex, and for monitoring purposes send sensory copies of the imitated actions back to the superior temporal cortex.

Play, Biochemistry and Emotions

A general neurological case has been made for how the idiosyncratic behavior of Latah could come about, but to complete the analysis the biochemical and hormonal functions must be explicated as well as more about emotions, especially anger. Panksepp (1998) makes the play connection writing about Tourette's syndrome,

> with its bizarre nervous impulses – which lead to tics and sudden verbal expletives, commonly including "forbidden" expression such as curses and slurs – may represent aberrant play impulses, or components of play impulses, circulating without restraint within the nervous system. Although this may seem far-fetched, pharmacological evidence provides some support for the hypothesis. Dopamine blocking agents, which presently are most effective in bringing Tourette's symptoms under control, are also very effective in reducing playfulness in animals. (p. 297)

Along with the host of numerous brain receptor sites, and neurochemical processes, the role of dopamine and neuropeptide transmitters most probably hold salience in the manifestation of playful behaviors. For example, Panksepp (1998) explains how the first discovered brain peptide, thyrotrophin releasing hormone, serves a peripheral body function in metabolic arousal, but a central role in the brain for arousal and by implication, play. This correspondent role of neurotransmitters is very common in mammals: So that insulin, for instance, in the periphery controls metabolic dispersion of nutrients, and in the brain helps mediate satiety.

Given the complexity of the brain and with at least several hundred different kinds of neuropeptides, and the pervasive effects of dompamine (as well as acetylcholine, aspartate, norepinephrine, glycine, glutamate, and gamma-aminobutyric acid) on everything the animal does, the precise mechanism of this facilitated play role is certainly not known at this time. "Also, descending glutamate systems from the whole neocortical mantle into the basal ganglia probably control every thought, every perception, and every emotional attribution that the brain can make" (Panksepp, 1998, p. 113). Therefore glutamate transmission has been a premier neurochemical system for understanding learning, memory, and consciousness.

More to the case of Latah, glycine receptors in the brain function in higher regions to control the intensity of glutamate transmission, and glycine, the simplest amino acid neurotransmitter, is directly linked to startle. This is powerfully confirmed when strychnine, a glycine antagonist, is given to an animal and consequentially startle is markedly increased (Panksepp, 1998). The startle reflex is the means that triggers the chemical, electrical and hormonal activation and elaborate response of the "RAGE system" (Panksepp, 1998). So immediately after the startle, cognitive inputs are transmitted largely via the amygdala's stria terminalis to the ventrolateral and medial hypothalamus and activation now certainly not related to, nor dependent on the higher brain areas, transpires probably within the dorsal half of the periaqueducted gray area of the midbrain with autonomic motoric and verbal responses.

Theoretically this makes sense, relative to the memory of being relatively demeaned in the Malaysian society: For over the years the accumulated reentrant learnings are transmitted into the system via thalamic and cortical inputs to the medial amygdala (Edelman, 1992). This is a different process than that caused by the more basic physiological irritations such as hunger and vital hormonal/sexual influences which enter the system via medial preoptic and hypothalamic inputs. Again it is to be remembered that the actual brain mechanisms that control anger are complex, are under multiple physiological, neuroanatomical, and neurochemical controls, and are only roughly understood. Panksepp (1998) states it definitively in that "the most important neuromodulators for the instigation of rage have not yet been definitely identified" (p. 203), and certainly distinct neurochemical systems are difficult to find for such higher, "secondary, cognitive-type emotions" (p. 301) with social context feelings related to shame, guilt, embarrassment, hate, contempt, and/or jealousy.

The playful behavioral quirks of the Latah and their sexual expletives also set the stage for what Brian Sutton-Smith (2003) theorizes about the significant relationship between the older reflexive and powerful emotional systems of anger, fear, shock and so forth and the newer systems for cognitive emotional survival including embarrassment, shame and guilt. So the excitement of play may lie in this resolution of the ancient and the modern characteristics, of the reptilian and the neo-homo ludic. Play, he writes, "could be the incentive for the genetic invention and selection of a method for mediating that conflict" (Sutton-Smith, 2003, p. 4).

Latah individuals may be acting upon many types of aggression from predatory to territorial and sex related to the types of maternal and irritable aggression which are not strong enough to provoke flight (Moyer, 1979). Emotional substrates like shame are also undoubtedly involved, so making a case that Latah has to be of one type or fit within any general typology is a mistake, for within the parts of the RAGE circuitry the various "forms of aggression are certainly not distinct at the subcortical level" (Panksepp, 1998, p. 193). Edelman (1992) makes the connection with emotions and thinking writing that "the range of human freedom is restricted by the inability of the individual to separate the consequences of thought and emotion" (p. 170). It is at this complex, unique, neurocognitive level that idiosyncratic behaviors are manifest, at the interface of the emotional cortical system and the genetically founded play system.

Although the literature does not confirm positive emotional feelings after a Latah 'attack,' it is very conceivable that this happens – albeit privately - in that if "the energized behavior of rage produces the desired changes in the environment, then it is rapidly mixed or associated with positive emotional feelings" (Panksepp, 1992, pgs. 195-196). During aggressive attack the type of available target is not as important as the fact that there is a living target upon which to find expression for one's rage. It's similar with monkeys who avoid the dominant ones and act out upon more submissive others. Tellingly, Howard (2000) suggests the following for constructively managing the emotion of anger: "Perform a productive physical task; Talk or act it out with a pet; Act it out with loud music, dance, shout; Play a physical game. ... Meditate in your own way" (p. 362).

SUMMARY

How does play relate to Latah's culturally specific performance pattern resulting from sudden startle leading to involuntary and normally inappropriate acts which may involve mimicry, obedience to commands, and/or the utterance of outrageous affect-laden words? The sexual expletives correspond to the 'dirty' nature of sex that which "is conducted in private, pondered obsessively, regulated by custom and taboo, the subject of gossip and teasing, and the trigger for jealous rage" (Pinker, 2002, p. 252). Perhaps Tricksters are a wonderfully historical example and Latah individuals are an active ludic syndrome example of not only the masking of the very centrality of emotions upon which play depends, as Sutton-Smith (2003) theorizes, but also of the social and personal playfulness that surrounds it. They serve as representative examples of the many playful culture-bound acts and ludic patterns that continue to camouflage the prevalence of and function of play in the world.

Individuals within groups and societies intuitively know the value and power of play. They naturally play it freely while other, so-called professionals, label it as abnormal and may call players "crazy." Leavitt (1993) argued about the nosological categorization of trance and possession disorder proposed for the 1993 Mental Disorders-IV Manual. He states correctly, in my view, that when trance and possession phenomena are positively valued in a society as with the case of *zar* or Latah, as opposed to negatively valued, the evident dissociation is not in itself an illness. Rather than treating it as a purely psychiatric disease entity, we need to consider what these trance or possession states may be expressing and what are the underlying playful substructures.

Relative to Latah and similar dramatic actions, Hildred Geertz argued that their frenzied manifestation could be understood only in the context of the courtly emotional restraint that is the cultural norm in Malaysia and Indonesia. That potential seems to be culturally universal, but more easily realized in Southeast Asian middle-aged or elderly women, who are already socially put-upon by societal messages. Latah therefore can be seen as "a parody or mirror image of Malayo-Indonesian values - and is 'funny' precisely for this reason" (Kenny, 1983, p.161).

Since Latah is the most widely written about culture-bound syndrome, could it be that the Malay Latahs are not only psychosomatically and nerochemically responding to strong hostile feelings against the cultural system they were born into, but also naturally responding to their anger against the many Western anthropologists, transcultural psychiatrists, and medical and psychological anthropologists who demean them ironically, often through their reporting of Latah? Alatas (1977) puts the prejudice well, stating that Malayan studies and "writing about latah as well as *amok* and other favorite colonialist topics created and perpetuated images of mental deficiency of the Malayan Other, which justified and encouraged European domination" (p. 177). Further, Winzeler (1995) confirms that most of the research on the Southeast Asian behavior of Latah "has been one of notes and articles, book-chapters and anthologies, rather than of monographs and books on particular instances" (p. 3). So-called culture-bound syndromes were explained through a synthesis of ethnographic and psychiatric or psychological filters and with conferences held and proceedings published and more instances 'discovered' and described a pattern was set (Caudill and Lin, 1969; Pfeiffer, 1968, 1982, 1994; Lebra, 1976). Therefore under these biased conditions a misunderstood natural hyperadrenergic survival response (startle) may be undergirded by play and not fully realized as an ironic humorous, ultimately tongue-in-cheek, disdainful social and culturally significant playful drama. After all the Malay prefer to limit the overt expression of strong emotion like anger and humor to theatrical and shamanistic performance.

The degree to which, if any, the imitative and quirky behavior of individuals with Latah affords them insight into their ludic pathology is not known, as is the extent to which the behavior is voluntary. One self-report of two Latah sisters is that they will "start imitating people for no reason" - at the same time! And unlike other Latahs, they will remember the event. Nonetheless a Latah is like "the prophetic trickster [who] points toward what is actually happening" (Hyde, 1998, p. 300), and serves the role of modern-day jester. Jesters told the truth in exaggerated, comic and masked form about societal concerns of the late medieval period, and Latah individuals may be dramatically revealing the truth of play's function and manifestation intrapersonally and within culture. Play works like that: The important lessons are not always obvious: like how the brain is pre-wired to relate to and hence play with other people and one's self.

Play is complex for it is biologically and socially given, involving order and disorder, display, entertainment, imitation, inversion, and marginality. At its center and central to human meaning are "the structures of imagination" (Johnson, 1987, p. 172). In its broadest sense, play actively reminds us that human action is many-leveled, and disciplines such as anthropology, physiology, psychology and even neuroscience cannot exclusively appropriate life events. Latah would not exist in the absence of the conjoined psychosocial, neuroelectrical, cortical and hormonal bodies of human players. Further this social, dynamic, exciting performance art with the underlying social psychological unease in the Malaysian culture fits the playing out of Latah. As with other worldwide ludic syndromes there seems to be a culture-specific neurocortical mixture facilitating a holistic bodily response that both masks and subverts the underlying causes in a playfully performed manner. The cross-cultural study of neurocortical mind/body/spirit process and psychosomatic dynamics lend themselves to a fuller explanation of the world of human playful power.

Rimbaud (1975) claimed that the only way an artist can arrive at life's truths is by experiencing every form of love, of suffering, of madness and to be prepared for it by a long immense planned disordering of all the senses. Given the multimodal organization of the social brain, maybe Latahs have a condition that is akin to synesthesia. At the very least the etomological foundation of the word is telling, for it comes from the Greek *syn* (together) plus *aisthanesthai* (to perceive). Synthesis is a similar word, as is symbol, from the Greek word *symballein*, "to throw together." "To perceive or throw together," what a wonderfully somatic way to realize what the brain/body/mind does. Play, particularly social play, aids survival and maintains a healthy mammalian organism, and at the same time actively confronts individuals with what it means to experience a lived body in process. To live fully is to allow foundational play, functional play, and phantasmagoric play to survive. Those individuals with Latah, like many of the world's synesthetes who have been gifted artists, composers and writers, through their art express a creative story of the truth, which reflects the complex nature of the social world and the very complex, ludic nature of the embodied human individual.

NOTE

1. The author is indebted to Brian Sutton-Smith for the creation of the term "ludic pathologies" and for help on an early draft of this chapter.

REFERENCES

Ackerman, D. (1990). *The natural history of the senses*. New York: Vintage Books.

Alatas, S. H. (1977). *The myth of the lazy native*. London: Frank Cass.

Bartholomew R. E. (1994). Disease, disorder, or deception? Latah as habit in a Malay extended family. *The Journal of Nervous and Mental Disease, 182*, 331-338.

Bateson, G. (1980). *Mind and nature: A necessary unity*. New York: Bantam Books.

Battaglia, D. (1993). At play in the fields (and borders) of the imaginary: Melanesian transformation of forgetting. *Cultural Anthropology, 8*(4): 430 - 442.

Beadle-Brown, J. (2004). Elicited imitation in children and adults with autism: The effect of different types of actions. *Journal of Applied Research in Intellectual Disabilities, 17,* 37-58.

Ben-Ze'ev, A. (2001). *The subtlety of emotions.* Cambridge: MIT Press.

Berardelli, A. (1995). Symptomatic or secondary basal ganglia diseases and tardive dyskinesias. *Current Opinion in Neurology, 8,* 320-322.

Budmana, C. L., Rockmoreb, L., Stokesc. J. & Sossinc, M. (2003). Clinical phenomenology of episodic rage in children with Tourette syndrome. *Journal of Psychosomatic Research, 55,* 59-65.

Buten, H. (2004). *Through the glass wall: Journeys into the closed-off world of the autistic.* New York: Bantam Dell Books.

Caudill, W. & Lin, T. (1969). *Mental health research in Asia and the Pacific.* Honolulu: East-West Center Press.

Damasio A. R., Grabowski, T. J, Bechara, A., et al. (2000). Subcortical and cortical brain activity during the feeling of self-generated emotions. *Nature Neuroscience, 3,* 1049-1056.

Damasio, A. R. (1994). *Descartes' Error.* New York: Grosset/Putnam.

Damasio, A. R. (1999). *The Feeling of What Happens.* New York: Harcourt Brace.

Drame, K. (1996). The trickster as triptch. In A. J. Arnold (Ed.), *Monsters, tricksters, and sacred cows: Animal tales and American identities* (pp. 230-254). Charlottesvile, Virginia: University Press of Virginia.

Edelman, G. M. (1992). *Bright air, brilliant fire: On the matter of mind.* New York: Basic Books.

Flanagan, J. R. & Johansson, R. S. (2003). Action plans used in action observation. *Nature, 424,* 769–771. August 14, 2003.

Geertz, H. (1968). Latah in Java: A theoretical paradox. *Indonesia, 3,* 93-104.

Hahn, R. A. (1995). *Sickness and healing: An anthropological perspective.* New Haven: Yale University Press.

Howard, P. J. (2000). *The owner's manual for the brain.* Atlanta, GA: Bard Press.

Hughes, C. C. (1985). Glossary of 'culture-bound' or psychiatric syndromes. In R. Simons and C. Hughes (Eds.), *The culture-bound syndromes* (pp. 469-505). Dordrecht: D. Reidel Publishing Company.

Hyde, L. (1998). *Trickster makes this world.* New York: Farrar, Straus and Giroux.

Johnson, M. (1987). *The body in the mind: The bodily basis of meaning, imagination, and reason.* Chicago, IL: The University of Chicago Press.

Kaufman, B. & Kaufman, F. (1995). *Son-Rise: The miracle continues.* New York: H J Kramer.

Kenny, M. G. (1978). Latah: The symbolism of a putative mental disorder. *Culture, Medicine and Psychiatry, 2,* 209-231.

Kenny, M. G. (1983). Paradox lost: The latah problem revisited. *The Journal of Nervous and Mental Disease, 171*(3): 159-167.

Kenny, M. G. (1990). Latah: The logic of fear. In W. J. Karim (Ed.), *The emotions of culture: A Malay perspective* (pp. 123-141). Singapore: Oxford University Press.

Kohler, E., Keysers, C., Umilta`, M. A., Fogassi, L., Gallese, V. & Rizzolatti, G. (2000). Hearing sounds, understanding actions: Action representation in mirror neurons. *Science, 297,* 246-248.

Landis, C. & Hunt, W. A. (1939). *The startle pattern.* New York: Farrar & Rhinehart Inc.

Leavitt, J. (1993). Are trance and possession disorders? *Transcultural Psychiatric Research Review, 30*(1), 51-57.

Lebra, W. (1976). *Culture-bound syndromes: Ethnopsychiatry and alternative therapies.* Vol. 4. Honolulu: The University Press of Hawaii.

Lindstrom, L. (1990). *Knowledge and power in a South Pacific society.* Washington, DC: Smithsonian Institution Press.

Llinás, R. R. (2001). *I of the votex: From neurons to self.* Cambridge, MA: MIT Press.

Mead, M., (1935). *Sex and temperament in three primitive societies.* New York: William Morrow and Company.

Moyer, K. E. (1979). *The psychobiology of aggression.* New York: Harper and Row.

Murphy, H. B. M. (1976). Notes for a theory on latah. In W. Lebra (Ed.), *Culture-bound syndromes, ethnopsychiary, and alternate therapies: Vol. 4* (pp. 3-21). Honolulu: The University Press of Hawaii.

Olanow, C. W. & and Tatton, W. G. (1999). Etiology and pathogenesis of Parkinson's disease. *Annual Review of Neuroscience, 22,* 123-144.

Omdahl, B. L. (1995). *Cognitive appraisal, emotion, and empathy.* Mahwah, NJ: Lawrence Erlbaum.

Osborne, L. (2001). Regional disturbances. *The New York Times Magazine,* May 6, 2001. Section 6, 98-102.

Panksepp, J. (1998). *Affective neuroscience: The foundations of human and animal emotions.* New York: Oxford University Press.

Pfeiffer, W. (1968). New research findings on latah. *Transcultural Psychiatric Research, 5,* 34-38.

Pfeiffer, W. (1982). Culture-bound syndromes. In I. Al-Issa (Ed.), *Culture and psychopathology* (pp. 201-218). Baltimore: University Park press.

Pfeiffer, W. (1994). *Transkulturelle psychiatric.* Second Edition. Stuttgart: George Thieme Verlag.

Pinker, S. (2002). *The blank slate.* New York: Viking.

Preston, S. D. & De Waal, F. B. M. (2002). Empathy: Its ultimate and proximate bases. *Behavioral and Brain Sciences, 25,* 1-72.

Prince, R. H. (1985). The concept of culture-bound syndromes: Anorexia nervosa and brain-fag. *Social Science & Medicine, 21*(2):197-203.

Prince, R. H. (2000). Transcultural psychiatry: Personal experiences and Canadian perspectives. *Canadian Journal of Psychiatry, 45,* 431–437.

Ramachandran, V. (2000). Mirror neurons and imitation learning as the driving force behind "the great leap forward" in human evolution. *EDGE*. 69: June 1, 2000. http://www.edge.org/documents/archive/edge69.html.

Rimbaud, A. (1975). *Complete Works*. Trans. Paul Schmidt. New York: Harper.

Rizzolatti, G. & Arbib, M. A. (1998). Language within our grasp. *Trends in Neurosciences, 21*(5), 188-194.

Saba, P. R., Dastur, K., Keshavan, M. S. & Katerji, M. A. (1998). Obsessive-compulsive disorder, Tourette's syndrome, and the basal ganglia pathology on MRI. *Journal of Neuropsychiatry Clinical Neuroscience, 10,* 116-117.

Schechner, R. (1988) Playing. *Play & Culture, 1,* 1 3-19.

Simons, R. C. (1980). The resolution of the latah paradox. *Journal of Nervous and Mental Disease, 168*, 195-206.

Simons, R. C. (1983). Latah II -problems with a purely symbolic interpretation. *Journal of Nervous and Mental Disease, 171*, 168-175.

Simons, R. C. (1996). *Boo! Culture, experience and the startle reflex*. Oxford, New York: Oxford University Press.

Stafford-Clark, D. (1971). *What Freud really said*. New York: Schocken.

Sutton-Smith, B. (2003). Play as a parody of emotional vulnerability. In D. E. Lytle (Ed.), *Play and educational theory and practice*. Play & Culture Studies (pp. 3-17). Westport, CT: Praeger.

Tourette, G. de la. (1884). Jumping, lata, myriachit. *Arch Neurol (Paris) 8*, 68-71.

Turner, V. (1967). *The forest of symbols: Aspects of Ndembu ritual*. Ithaca, New York: Cornell University Press.

Winzeler, R. L. (1995). *Latah in Southeast Asia: The history and ethnography of a culture-bound syndrome*. Cambridge: University Press.

Yap, P-M. (1969). The culture-bound reactive syndromes. In W. Caudill & T-Y. Lin (Eds.), *Mental health research in Asia and the Pacific* (pp. 33–53). Honolulu, Hawaii: East-West Center Press.

Part III

The Play Systemics of Education

If there is some ambiguity about the adult "systemic" ludic intrusion in some of the above examples, there is none in those of the present section. Here we have quite extensive efforts to use play as a means of advancing the school moral or curriculum values. Here we have the rhetoric of play as a form of progress, the most extreme example being Trageton's account from Norway.

CHAPTER 8. ARNE TRAGETON; PLAY IN LOWER PRIMARY SCHOOL IN NORWAY

While Trageton's Norwegian "play intrusive" efforts are similar to many others tried throughout the Western world, it is doubtful if there has been many such comprehensive State attempts to change the whole nature of the elementary school day to make it become as far as possible a child player's laboratory rather than a teacher's lecture place. And what one sees so clearly throughout the multiple examples that are provided here is that re-educating the teachers is the most difficult hurdle to overcome. What is also apparent is that it is difficult to make teachers realize that the new play curriculum is more about the dramatics that can be induced into learning than it is about the earlier types of rule bound games that have been made a part of some secondary school curriculums. What has to be introduced is that education is to be an open ended social drama, not just a series of tight rule bound games such as one finds in the earlier literature as in Clark Abt's "Serious Games" (1970), valuable as they are on other grounds. In Trageton's terms, the whole school process is to be kind of culture creation, and the classroom is to be workshop where that process goes on. Consistently he also discusses turning the outdoor playground into a similarly inventive place which is what many playground designers have tried over recent decades, in some cases even turning playgrounds into gardens (Moore, 1986). From the present editors' point of view, all of these difficult efforts are parts of a larger global trend to emancipate the human imagination from the stereotypic and even barbaric confinements that have served it so well in times past. The task is perhaps to allow the imagination the kind of variants that will heighten the possibilities of progressive human evolution. But perhaps a further and more important point in favor of Trageton's attempts is that in a world of increasing globalization we need a humanity that is more flexible in ways of constructing social and intellectual realities at all levels and on many occasions. And the school is the place for this to begin in co-operative dramatizations. What we have inherited culturally unfortunately is rather a multiplicity of orthodoxies of religious and political kinds which have been useful in creating larger national polities than was the case at prior foraging levels; but are now themselves a hindrance to any jump forward in global humanistic integration.

CHAPTER 9. LYNN ROMEO AND KRISTIE L. ANDREWS: THE MILL LAKE HOSPITAL: PLAY IN A MIDDLE GRADE CLASSROOM

The educational commitment to play in this study is more typical of con-temporary work in the USA on the role that play can have in the stimulation of literacy (Roskos & Christie, 2000). The procedure here was to make pupil col-laboration a part of the exploration of the elements (protractors, journals, re-cords, etc.) in a pretend Hospital Center with focus on the study of the human body and its support systems. The authors were somewhat vexed by the amount of repetitiveness that took place in some of the subjects' own inquiries, not rec-ognizing perhaps that redundancy is a primary characteristics of most forms of social play (consider hopscotch or basketball). They were a little vexed also that the participants enjoyed the puzzle games that were provided over most other forms of activity. Still, there was high pupil engagement and, high self-regulation that the investigators judged positively. It hard to say, however, whether what is going on here is a kind of democracy of learning seriousness or a testament to the value of play oriented methods. In some ways the supervision and controls exercised by the teachers who were nevertheless positive to child participation and inventiveness reminds one of the culture play systemic amongst the children of Morocco in Rossie's research (See chapter 5 above).

CHAPTER 10. ANNICA LÖFDAHL: PRESCHOOL TEACHER'S CONCEPTIONS OF CHILDREN'S "CHAOTIC PLAY"

It is something of a relief in this chapter to know that we are again in the presence of a very real child play form. The hybrid forms of the prior two chap-ters, while important in their own right, are only playful out of sufferance. It is a relief also to realize just how difficult real child play can be for the adults who have to manage it. Anyone who has been a teacher of preschoolers will know just how riotous and at times dangerous the children's own play can become. It is not unlike the locomotor play of young animals which can proceed back and forth quite precipitately often with one animal leading the others and then sud-denly turning on them and having them flee in the other direction. In play-grounds all of this can proceed with some distant supervision unless it truly en-dangers itself (which is seldom). Indoors there is obviously less space and the teachers' intrusions with useful fantasies make good sense.

Chapter 8

Play in Lower Primary School in Norway

Arne Trageton

Norway lowered the school entry age from 7 years to 6 years in 1997 and the new National Curriculum (1997) stipulates that play should have an important place, not only for the 6 year olds, but for the entire lower primary school (6-10 years). Cross subject teaching should dominate through long run themes in a playful learning process. The last 12 years in Norway have also shown a massive expansion of a leisure time institution at school before and after compulsory lessons, called SFO. In 2000 about 50% of all school children 6-10 year olds were using SFO in the early morning and from about 1 to 5 pm. About 2/3 of the total time at SFO is reserved "free play." However, Norway had little research about play among 6-10 year olds in educational institutions. Also internationally, there were only a few research projects about play in primary schools (Retter, 1983; Hartmann, 1988; Hall & Abbott, 1991; Moyles, 1989, 1994; Christie, 1995, 1999; Pessanha, 1995; Wassermann, 2000). This situation was the background for my research project.

THE NEW NATIONAL CURRICULUM

New Demands of Teacher Training

Since 1997 Norway has adopted the most "playful" National Curriculum in Europe for 6-10 year olds. What about the realization of this revolution? Preschool teachers were invited to move from preschool institutions to work in lower primary school in grade 1, and if they enrolled in additional one-year education in initial teacher training, they were qualified to teach all children ages 6-10 years old. This was a revolution for the preschool teacher. In the new grade 1, one preschool teacher and one primary school teacher cooperate. For the primary school teachers the reform was also a revolution. They had hardly heard about play in their initial teacher training, and for them there was unfortunately, no demand to enroll in further education for the new play pedagogy. However, many of them found that they were unqualified and voluntarily took the same course as the preschool teachers. Since 1986, the University of Trondheim has taught these courses to the top 160 students all at one time. Now most of the teacher colleges in Norway have similar courses. The need for play research and textbooks in the area was great.

My research project from 1994 to 1997 provided a description and interpretation of 6-10 year old role play and drama with the help of video documentation in six different school environments. The environments varied from child controlled "free play" in SFO to adult controlled structured playful teaching. Based on 50-60 hours of video recording, this provided the background to produce teaching aids consisting of a series of video examples and a textbook to help teachers for their new role (Trageton, 1997).

Play Research and Teaching Aids

Most textbooks for students are structured for preschool teacher education. Yet Johnson et al. (1999) for instance, have only a few words about play in the elementary grades. A similar textbook structure is useful also for primary school teacher education. Because of the strong position of the National Curriculum and 10 different school subjects, it seems natural to start to discuss the relation between the National Curriculum and play. After discussing the position of play in relation to the written curriculum, the teacher-educator needs to discuss definitions of learning and play and what play forms are most important and valuable in lower primary school. Because rule play has dominated the school tradition and discussion, it seems appropriate to raise the question if dramatic play is more complicated and important at that age than older ages. The teacher's tasks and physical frame factors are central for stimulating play, and it seems important that the student teachers make their own small research project to improve practice.

Play in the New National Curriculum

The debate about lowering the school entry age to 6 years started in the early 1980's. The argument for this was to harmonize the school entry age with that of most countries in Europe, and to fulfil the belief that starting the academic learning process earlier would be effective. The argument against this was that preschool for 1-7 year olds still had the best pedagogy with its dominating play curriculum. From 1986-1990, Norway completed a large-scale pedagogical experiment with 6-year-olds in 42 communities in three different settings which were:

a)   6-year-groups in preschool

b)   6-year-groups that came from preschool to primary school for a day or two per week for some pedagogical programs together with the 7-year-olds at school.

c) 6-year-groups located at school in a separate classroom or building.

As expected, the "a" program gave the children more play than the "c" program. During the four years, free play was reduced from 51% in "a" to 23% of the time in "c." Teacher-structured programs were doubled in the same period (Haug, 1991).

In the 1990's the debate continued until Parliament made the decision to extend the compulsory school age from 9 to 10, with 6-year-olds added as the extra year at the bottom. At that time there was strong resistance in the country against lowering the entry age out of fear for a less playful pedagogy in primary school than heretofore in preschool. Because of that and also because three of the political parties were against lowering the school starting age, the political compromise was that the revised lower primary school (6-10 year-olds) should be formed from "the best of the preschool and the best of the school program." And the "best" from preschool referred to "play" of course, with the first year at school being as similar to preschool as possible. Parliament demanded that the teaching program in lower primary school have a definitive preschool form. The school days should mainly be organized around children's need for play and free time activities. Play is a goal in itself for personality development (free play), but the teacher should also use play as a working method for teaching cross-subject themes (structured play). Also to movtivate children in the different subjects playful teaching was encouraged. Learning through play was the slogan. The majority in Parliament meant that perhaps the most important new school reform component was that more play should be allotted for the 6-10-year-olds in school. The National Curriculum has followed up on what Parliament mandated on play. Now Norway has the most "playful" plan in Europe for 6-10-year-olds. The subjects have the following time allotments for the first four years at school:

Norwegian language	912 hours
Mathematics	532
Religion	266
Play	**247**
Physical education	228
Art and craft	228
Social science	190
Science	152
English	95
Home economics	38

For the first time in history, "free play" is quantified on the timetable. It is the fourth greatest "subject" in lower primary school!

Another revolution is that cross subject themes shall dominate the learning process at least 60 percent of the total time. So 60 percent of all the lessons in the respective subjects used for a thematic approach to teaching. The plan predicts that a theme-based approach should inspire a child to play, and moreover, play should bring inspiration to the themes. The working methods in lower primary school are creative activity, play and practical work, with the play method being the dominating one. The Norwegian language curriculum describes the importance of role play and drama; mathematics favor play and game; religion incorporates dramatizing the stories; art and craft stresses constructive play; and physical education focuses on sensorimotoric and physical play with rules.

Learning through play

Different definitions of learning reflect different scientific paradigms. They also reflect what learning in school means. Bjørgen (1992) defined learning as seeking meaning and similarity relations between new knowledge and existing structures based on experience in the learner. In such a constructivist definition the responsibility of learning is put on the learner, not on the teacher. Bjørgen placed different learning forms into the following hierarchy:

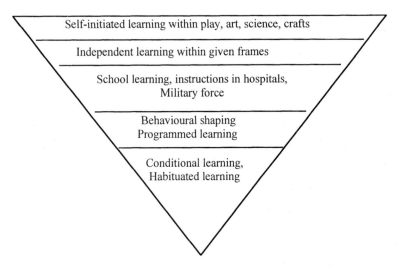

Bjørgen (1992) suggests that behaviorism is characterized by learning at the lowest level. Here the teacher is primary, not the learner. This is an amputated learning. The highest level in the hierarchy is where the learner has full control. It is interesting that for Bjørgen, play and scientific learning are at the same level. This theory is a strong argument in favor of play as fundamental for child controlled learning. Bjørgen's definition is in harmony with the learning concept in the new National Curriculum.

Play definitions from Comenius to post-modernism

When play became important in school, play theories became important. Trageton (1997) provided a summary of play theories from 1632 to the 1990's in a compressed, popularized form. All these attempts say something essential about play, but also characterize the time and research tradition behind the definition. This is similar to what Sutton-Smith (1997) called "the rhetoric of play." Simplified versions of the short definitions are:

Play is action, symbol making, work, edifying (in religious meaning) (Comenius, 1632)
Play is children's art with synthesis and emotions/cognitions (Schiller, 1793)
Play is the highest point of development in childhood (Frøbel, 1826)
Play is surplus energy (Spencer, 1855)
Play is practice training for adult life (Groos, 1899)
Play is emotional outlet (Freud, 1920)
Play is expression and self-mastery (Eriksson, 1941)
Play is assimilation of the world, one condition for cognitive growth (Piaget, 1962)
Play develops abstract thinking (Vygotsky, 1933)
Play is process, creativity, flexibility (Sutton-Smith, 1971, Bruner, 1972)
Play is group-adapting, reflection of reality (Leontiev, 1977)
Play is work (Goodman, 1974)
Play is fantastics, aesthetics (Rodari, 1991)
Play is communication (Bateson, 1955; Garvey, 1979; Olofsson, 1987)
Play is ecological cooperation, interaction (Bae, 1988)
Play is culture teaching (Opie & Opie, 1969)
Play is culture creation (Huizinga, 1938; Gadamer, 1965; Sutton-Smith, 1990)
Play is *spiel* (Wittgenstein, 1932; Papert, 1983; Rasmussen, 1990; Bauer, 1996)

In my opinion the last two definitions are the most representative for the rhetoric of our time, and therefore perhaps most useful for us in our post-modern world. Therefore in my book I focus on the last two definitions, although included in the book is an attempt to demonstrate that in the analysis of all sorts of play among the 6-10 year-olds the teacher may have to use a variety of these definitions to realize and describe what is going on. For preschool teachers in Norway I try to expand their knowledge to include more general theories than the traditional preschool theories.

Play forms in lower primary school

What type of play forms are most useful for the 6-10 year-olds? Although different play theorists use different names and categories, they roughly group into four play forms:
a) Functional play, psychomotor play, practice play, introduction play;
b) Role play, sociodramatic play, make believe;
c) Constructional play, play with materials, building play; and
d) Rule play

Rule play is, according to Piaget (1962), a more advanced form and should therefore be the most common play form for 6-10 year-olds. Most play researchers of this age group, and the teachers in lower primary school in Norway seem to have accepted Piaget's theory and have traditionally used this play form, suggesting that others are childish and relegated to preschool children. However, Sutton-Smith (1971) and Gardner (1983) do not agree with Piaget that rule play is the most advanced play form. Dramatic play can have more complex and intellectually more demanding and flexible rules with sudden shifts from time to time. The children have to use all the intelligences, which Gardner describes, not only the logical mathematical and verbal intelligences as suggested by Piaget. In Germany, Otto (1990) found that children 6-8 years old in *Grundschule* (primary school) were more interested in dramatic role play than in rule play. However, the teachers and parents follow Piaget's theory, expected and organize rule play. They arrange the physical environment to include different ball games; hopscotch; and other games in leisure time; games in mathematics; rhythmical rule play in language and music; and so on. In addition, the adults accept and stimulate the children in this type of play. Dramatic play was hardly expected, and neither the physical environment nor the teacher assistance stimulated this sort of play. Observations from 1994-1997 (Trageton, 1997) documented with 40-50 hours of video recording supported the observations of Otto, and the criticism of Piaget by Sutton-Smith and Gardner. The dramatic role play for 6-10 year-olds seems to have much more complicated rules than the rule play by the same children, *if* the school encourages dramatic play. Reorganizing the physical space proved to be very important. Reorganizing the timetable also was necessary. Courses designed to focus more on dramatic play, training in observation of play as well as an active adult role in relation to dramatic play (Christie, 1992) was important to change attitudes about what sort of play was important in school. The relation between the play forms represented in Figure 8.1 shows a structure other than Piaget's theory.

Instead of saying that rule play is the most advanced, this model suggests that with a common root in sensorimotor (sensorimotoric) play one can see three different branches of play in primary school. I think it is important not to forget the primary fundament in all play forms. The child always uses the senses and motor activity in play. The arrows show the development outwards. First, the children develop divergent, flexible symbols in constructional play (Trageton, 1994a) and in dramatic play (Heggstad, Knudsen & Trageton, 1994). This corresponds with the four-year-old children in preschool. Many 6-10 year olds are at the same level, *if they are not trained in constructional play* (Trageton, 1994b). The most common level for 6-10 year olds in both constructional play and dramatic play is play dominated by convergent thinking - on how to build a special oil platform in the correct representation for example. Another example would be the proper, systematic realization of role play about a special adventure, family life and so on. It seems to be a good hypothesis that also in rule play, creative, flexible use of the rules will be an earlier stage than the convergent, logical, systematic rules all children have to accept at a later stage.

Figure 8.1
Play Forms In Primary School

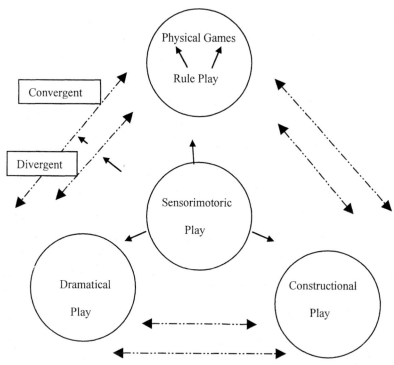

Rule play has many good descriptions in earlier books in Norway. Constructional play as a base for workshop pedagogy is described in Trageton (1994b). Therefore the focus is dominated by dramatic role-play, which has had a minor place in Norwegian schools up to now.

Preschool 6-year-old play principles

Because the 6-year-olds in Norway until 1997 have been preschool children and play filled 50-60% of the time, schools can learn a lot from how preschools have organized play for 6-year- olds. What consequences will these experiences have for the new grade 1 at school? Learning and understanding of the constructional play and dramatic play tradition in preschool is important for the primary school teacher. Additionally there is a lot of material from earlier textbooks about play in preschool. The most important lessons to be learned from preschool education are the following:

- play groups should be small
- the classroom should be divided into small corners for playing
- classrooms should look like workshops

- classrooms should be close to outdoor areas
- outdoor space should look like workshops
- inspiring and varied materials are necessary to create different forms and environment
- adults should have an active role, through observation, dialogue and involvement in play
- boys' play, girls' play and unisex play should all be stimulated
- frame play should be planned (organizing milieu to link several play-groups together in complex play within a common theme)

Role play and drama for 6-10 year olds

The new National Curriculum (1997) describes play in three different versions: play as free development; play as a working method in the thematic cross-subject teaching; and playful teaching within the different subjects to further motivation. This was the foundation of the 1994-1997 research project (Trageton, 1995, 1997). The video recordings throughout the project period were done in six different settings within this frame:

Play as development (child controlled, free play)
  a) Free play in SFO (leisure time after school)
  b) Free play in school lessons
Play as working method (learning centered play)
  c) Drama/role play in workshop pedagogy
  d) Drama as free choice in SFO
  e) Drama as method in thematic teaching
Playful teaching in different subjects (adult-controlled)
  f) Mathematics, music, language

"Play" in the Norwegian National Curriculum spans from the most child-controlled to the most teacher-controlled "play" or drama. It is important to show examples from the whole range of this "play" concept in the Norwegian school.

Play as development

a) "Free play" in SFO (leisure time institution at school)
Haug (1996) found that 6-year-olds in Norway were playing 69% of the total time in SFO. The adult role was usually passive. The argument is that children should have free time to do what they want, most of their time. They can play where they want, with whom, with what, and when they want during the SFO period. This often leads to stereotype repetition of earlier preschool play. The personnel have little knowledge about play in the 6-10 year-group (Lidén, 1994). In my project the children in one SFO were followed for a year with a video camera. The focus was on dramatic play, not rule play or construction play. Of 40 hours of video material, 18 play observations lasting for an hour or longer were chosen, and through editing compressed to 3-8 minute play

sequences. The videocassette with 18 play episodes should be used in teacher training to analyze and discuss the quality of the play from a motor, social, emotional, aesthetic, cognitive and communicative point of view. Also the impact of the physical organizing for play and the adult role can be recognized and discussed. I give a brief description of some of these episodes in the book, to show typical play episodes and the variation of the different themes in play in one SFO.

b) Free play in school lessons time

In the new National Curriculum, 247 hours during the first four years at school should be "free play," most in the first year. This is quite new and a severe demand for the teachers in primary school. A class of 7-8 year olds were followed for two years, where two playful teachers dared to use 3 hours of free play each week, two years before this reform started. This experience gave us useful information about an educational area all the classes shall do after 1997. The classes had one hour "free playtime" two days per week before lunch. These classes were trained in group work and workshop pedagogy, and therefore it was rather easy to arrange ten different play corners. The children chose in which corner they wanted to play, but if three or four children had chosen one play corner, no more children could choose that corner. This is not unlike the "High scope" system in the United States for 3-6 year olds (Hohman, Banet &Weikart, 1979). Later on the classes also used the outdoor area for these periods. The development in this play over two years is documented through edited videos, and the textbook (Trageton, 1997) which gives a verbal description of this development. Paradoxically, there seemed to be a higher level of play within this more restricted time and space compared to SFO. Perhaps the narrow time schedule, a systematic choice of play corner, and a more active adult role inspired the children to use more energy and commitment in play?

Play as working method

c) Drama/role play in Workshop pedagogy

Workshop pedagogy (Trageton, 1994b) is a way of organizing teaching lessons for 6-10 year-olds inspired by the Laboratory School in Chicago (Dewey, 1900) and the British primary school integrated day (Brown & Precious, 1969). The movement started in three classrooms in 1978 and gradually spread across the whole of Norway. Each class is divided into 5 groups working with different art and craft materials: clay, sand, blocks, rigid materials and flexible materials. The children work within a common theme across the subjects, for instance "our town." Through structured construction play the children make representations of different sides of the theme including houses, cars, people and so on. After the constructing, they dramatize what might happen in their structured environment. Then from their experiences, they write a story about their special theme, and at last, they compute mathematics around what they have made. The main role for the teacher is to be a dialogue partner to expand concepts and language around the concepts in which each child is interested. This strategy from concrete to abstract has for the last 20 years proved effective both for slow and fast learners in the classroom in Norway (Trageton, 1994b). As mentioned in the beginning of this chapter, the National Curriculum pronounces that the dominating working methods for 6-10 year-olds should be creative activity, play and practical work. The workshop pedagogy is in tune with such a demand. The textbook (Trageton, 1997) describes the dramatizing stage in workshop pedagogy. A class was followed for 2 years (7-8 year-olds). Ninety-five video recordings were made of the drama stage. Twenty-four of these episodes were chosen and edited on a videocassette as examples of the development over two years. The textbook describes the development in detail. In the beginning, the dramatization was only actions. Gradually one child in the group played two roles, later two and three children played together, but it was very hard to get real cooperation between all the children in the group. The rest of the class was the audience.

In another school, the teacher documented the effect on written language. From a beginning of 10-12 written words, the texts grew rapidly to an average of 80 words in the old grade 1 (7 year olds). The class performance ranged from 12 to 310 written words. To provide an impression of an advanced level let us take a story from the theme called Skudesneshavn (a tiny town where the children live). The children had made some houses in town for family members who wanted to buy the houses. In the dramatizations, the figures representing husband and wife went into the house to take a look. They tried to go through the door, but the door opening was too narrow for the figures. They improvised and went down the chimney instead. This resulted in much laughter. One of the stories composed afterwards was:

> Once upon a time a man and his wife wanted to move to another house. But they did not know that they had ordered such a small house. They had to creep in the door. But the man had a good idea; he said that they should try the chimney. But the chimney was narrow, so it was hard to get through, so they wanted to find another house. They went to the man who owned the biggest house in town. But he was a real miser, and he said no. So they began to search for another house. At last some-body showed the way to a house of just the proper size.

d) Drama as free choice in SFO

Among six different courses consisting of 2 hours a week for 12 weeks, the 8-9 year-olds could choose a drama course. This drama course was very much structured by the teacher. The children dramatized a family who travelled from Norway to the United States in the 1880s. This family was followed in 12 episodes: first the farewell at home, then the sailing trip, the immigration problems in New York, finding a farm in the Midwest, fighting with Indians, inventing the steam engine, using a spinning machine, and in the tenth episode, moving to Chicago to produce cars. All the time the teacher practiced the teacher-in-role principle. She dramatized and let the children dramatize and role play the proper roles within these narrow frames. The groups built their creative, original cars from old school tables and chairs, cardboard boxes and other scrap.

Then the teacher became a director who wanted to give a prize to the most original car. Afterwards there was a car race, where notably the boys showed a deep commitment in their playing. According to the new National Curriculum (1997), this playful dramatizing could also have been a good type of teaching relative to science, social science, language, mathematics, art and craft. Is this dramatizing real play? According to the National Curriculum this is play as working method.

e) Drama as method in thematic cross-subject teaching

The new National Curriculum says that drama should have a strong position as working method among 6-10 year-olds. A video documentation from the cross-subject theme of Stone Age in grade 3 (9-10 year-olds) gives a glimpse of a theme lasting for four weeks. The drama teacher together with the classroom teacher used the teacher-in-role technique. The drama teacher was the Stone Age leader, and the children acted as the tribe who went to make stone weapons in the little forest outside of the school. Then the classroom teacher became a representative for a Bronze Age king, and the adults created a conflict where the children have to choose if they wanted to become slaves under the Bronze Age king and learn to make bronze swords and jewellery, or fight the bronze warriors, eventually fleeing back to the Stone Age. The children themselves had to solve this conflict, because now the Stone Age leader (the drama teacher) suddenly became old and sick, so he could not give them any advice. By systematically playing Stone Age roles and Bronze Age roles, combined with reading about these ages in books, making stories around the theme, making stone carvings, paintings and so on, this drama or structured play as working method made the historical material more attractive and engaging for the children.

f) Playful teaching in different subjects

The Norwegian national curriculum uses this formula with the argument that although this is not play, in the ordinary sense, a "playful" teacher can use "playful" methods to motivate children for specific learning activities. Here my examples were borrowed from my students taking further education in 6-10 year olds pedagogy. One student organized a shop in the classroom corner to teach the children the numbers from 1-9 and the addition of these numbers. All the prices in the shop were from 1-9 Norwegian *crones*, and the children played shopping in a structured manner, and took notes of the results of the shopping in their mathematic books. Another example was the music lesson where the teacher wanted to teach the children a classical song, while the 8-9-year children were most interested in heavy metal music. The result was a playful combination of the classical melody, where the children played that they were members of a heavy metal band, and with the help of the teacher, composed the rhythm and beat they found suitable to "jazz up" up the ancient song. A third example was when the grade 1 children played with personal computers and text programs to make texts. The teacher "played back" comments, and so the children could make a new and better version of their text. This was process oriented writing in grade 1.

FINAL COMMENTS

The danger of expanding the play definition as far as to "playful teaching" is that teachers can say that by referring to "playful teaching" they cover the total demands about play in school. I try to show the readers of the textbook that according to the National Curriculum the teachers have to use the entire range of the play concept. Classroom teachers have to train themselves especially to accept and master "free play" at school. Preschool teachers, on the other hand, have to accept that play, as working method and playful teaching, should also have its place inside the play concept at school.

The teacher's tasks in play

The main tasks were play observations and the different adult roles.
a) Play observations.

While preschool teachers have a long tradition of play observation, primary school teachers in Norway have not. To train teachers to make play observations as a new routine at *primary* school is a radical new task. This chapter tries to offer strategies for very simple descriptive/interpretive observations, short daily logs, and spontaneous reflections. For busy teachers I think it is better to train them to make observations and short logs as a simple daily routine than to teach them such complicated observation forms that the average teacher finds no time to make such observations. The most important thing is to train the teacher to draw consequences from the observations for structuring the total learning climate in the classroom. As mentioned earlier, we have no tradition of play observations in Norwegian schools. However, the play observation tradition in English primary schools (5-11 year olds) provides good examples of observation schemes (Moyles, 1994). This book was translated into Norwegian in 1996.
b) Adult roles in play

In my opinion a strict routine about observation of play is the best way to change the ordinary teachers' attitude toward play and different adult roles in play. The classifications by Wood, McMahon, and Cranstoun (1980) and Christie´s scale from ignoring to dominating roles in play are a good start (Christie, 1992). In addition, the adult roles of Hartmann (1988) are useful. The most important thing is that the teacher not only learn the different adult roles to stimulate higher quality in play, but practice these roles (Enz & Christie, 1997). Only then is a better understanding of the play possible.

Physical frame factors

School building, classroom and furniture placement, outdoor arrangement and organization of time are important factors for play development. The question is whether we want an auditory school or a laboratory school. Norwegian school history shows that the auditory school with traditional classrooms and students listening to the teacher's lecture still has strong support, in spite of the National Curriculum.

The preschool in Norway and *"The Integrated Day in Primary School"* (Brown & Precious, 1969) in England are much more like a laboratory school ideal with children's concrete activities in the center. Figure 8.2 shows such a classroom.

Figure 8.2
Norwegian and British Preschool Model

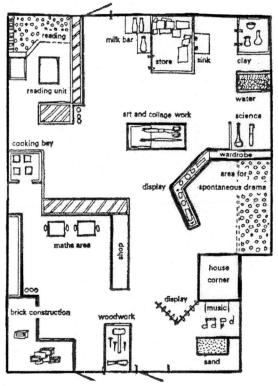

When the new National Curriculum suggests that creative activity, play and practical work should be the *dominating* learning methods for 6-10 year-olds, the school and the classrooms should be more laboratories than auditoriums. Because six year-olds until 1997 have been in preschool, it has been widely accepted that their classroom ought to be quite similar to a preschool room. The problem is to convince all teachers that this would be a good classroom also for 7-10 year-olds, instead of a more auditory setting. Figure 8.3 shows the structure of two classrooms for 7-8-year olds, where the video recordings of "free play" in school lessons were done.

Figure 8.3
Norwegian Classrooms

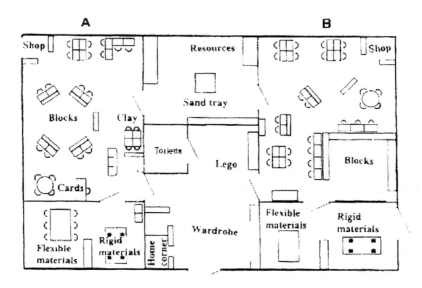

The classrooms, labelled A and B, are constructed for small groups of children doing creative activity, play and practical work. The classrooms and group rooms are constructed for workshop pedagogy where small groups of children are working with clay, sand, blocks, and rigid and flexible materials to construct different concepts in relation to the cross-subject theme going on in any particular month. The groups also form the bases for other practical work and their workbooks in different subjects. For "free play" lessons mentioned earlier, children use the different corners as play opportunities from which to choose. Here they also can choose among the three corners in the wardrobe. Class A may use the extra play area on Monday, and class B may use it on Tuesday. To avoid pedagogical barriers between the 6-year-class with preschool teachers, and the 7-year-class with primary school teachers, an ideal structure would be to have the 6-year-olds in classroom A and the 7-year-olds in classroom B. Also a multi-age pedagogical program for both classes and teachers a few hours a week is recommended. For instance, in the "free play" periods and in workshop pedagogy a common theme for the two age groups could be provided. This is now a common structure in many places in Norway.

What about the auditory function of the classroom? The block corner is so big that the teacher can gather all of the 20-28 children around her to tell a story, sing a song, or to discuss a common plan for what to do in the workshops that day, or to evaluate the work the children have done the previous week. This textbook (Trageton, 1997) gives a lot of advice about the furniture, materials and tools for learning within the different subjects.

Planning the Outdoor Area

In Norway outdoor activities have held a high status. Norwegian children have a long tradition of outdoor sport activities and play. However, in recent years, children in Norway have spent more and more time indoors. The play possibilities outdoors in forests and open fields has been greatly reduced: Children sit more often in front of televisions, videos, radios and computers at home. Because the school now expands the time radically for 6-10-year-olds, the tendency for indoor life is stronger. Therefore, there is a movement in Norway to stimulate "outdoor school," outdoor play and so on. Nevertheless, in practice the outdoor environment often is a black asphalt square. Here the schools follow Piagetian thinking that rule play, hopscotch and organized sports like football should dominate the play possibilities. Areas for construction play and dramatic play are almost nonexistent. It has become difficult to make the outdoor environment at school more attractive for 6-10-year-olds. What are the physical demands for better possibilities for outdoor play? Regarding the new reform, many communities and schools have bought play apparatus from commercial firms, like climbing trees, slides, swings, and carousels without reading all of the research reports, which show that this costly apparatus occupies only a few children for a short time, when it is new. Most research shows that what children really need is an outdoor space divided into small corners with the help of plants, walls, and so forth to shape a safe environment and framed corners for small play groups to organize their own play. A lot of unstructured play material is also necessary so they can form their own play environment.

The most important thing is to stimulate a "building playground," a Danish tradition we have tried to implement in many Norwegian schools during the last 12 years. The last example is one school, which planned an area for 30 huts for 150 children (6 and 7 year olds), 5 children per hut. The parents constructed the skeletons for the huts, and reserved the building area for the children. Afterwards the multi-age small groups built on their huts for months. Plain asphalt is a good base to build upon, because the "streets" then are ready beforehand. The children had to make a general plan, and figure out how to put 30 huts into that restricted area. Figure 8.4 shows one example of a hut with specifications and two town planning examples constructed within the 16.8 by 13.2 meter space (18.4 yards by 14.4 yards).

Figure 8.4.
Hut Construction and Town Planning

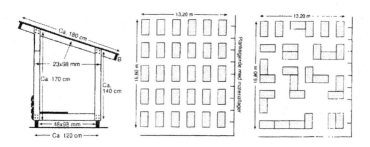

Cheap materials make the cost of this "town" minimal compared to the permanent, commercial play apparatus the children apparently don't use. Here 30 groups of children played both construction play and role play at the same time. According to the National Curriculum cross-subject thematic teaching should dominant for 6-10-year-olds. "Our town" would be a good theme for several months. Also social learning has realized positive results, and bullying at school has been reduced. Most school subjects could be integrated in that one theme. After three or four years the town would be deconstructed so the newcomers at school could build their own town. Much of the material can then be used for a second time.

To expand the traditional motor play to be more than football, many schools try to make use of real trees and a natural environment combined with ropes, climbing possibilities and so on. The third important area is the "learning through landscape" movement, inspired from England (Titman, 1994). This stimulates biological/ecological thinking in children. Greater emphasis on these qualities will provide better possibilities for children's play. Finally an example from the north of Norway underline the importance of snow as play material: Two students in the one year additional course in 6-10-year-pedagogy showed through their project work that in their climate with snow 6-7 months of the school year, there were many possibilities to play with that free material. This had not been developed before in a systematic way. For instance, they used the material in a similar way as in modern concrete buildings, where the children with the help of formwork could "mold concrete" with snow and make a perfect house with a pillar row in the front in 20 minutes. With snow as "concrete" they could strip the forms immediately. Afterwards the role play in the house started at once. Another group used a rectangular plastic bucket to form snow blocks to make a roman arch, and the same principle was used for igloos, barrel roofs, and gothic churches. The children also made huge snow sculptures, tunnels and holes. This project showed how to use the snow systematically to stimulate diverse playing over a long time, and not only as short "happenings" occasionally.

Local Innovations and Inspiration for New Ones

In the one year additional course in 6-10 year old pedagogy, students created a research experiment and innovation from a special area within the 6-10
year old free choice pedagogy involving language, mathematics, music, or
cross-subject themes. During the last 8-10 years, more and more students have
chosen something related to play in primary school. Until now 50-60 students
have made small projects documented in reports of 20-30 pages, some of which
have already been mentioned. Many of the projects have been most useful in the
textbook (Trageton, 1997), for they serve as inspiration for different schools to
start new play projects.

Does Play Pay? Experiences From Other Countries.

The revolution in Norway with the new National Curriculum mandating
that play now have a dominating place for the first four years in primary school,
is a political compromise. Conservative parties wanted to start academic training
at an earlier age. Some parties were against lowering the school start age. To get
political acceptation for earlier school start, the majority in Parliament decided
that the program for the lowest level should be dominated by play and cross-
subject teaching. Many people now ask anxiously: Does play pay? We still have
no answer, because we have not tried this program in practice for sufficient
years. I hope a major evaluation research program follows this radical pedagogical experiment to show to what degree this program is realized, and what effect
this has on academic learning. Because there is little Norwegian research about
play in school, experiences from other countries serve as positive examples.
England has since the 1960s a long tradition of informal education and
learning-centered play as the dominating method. Wassermann (2000) in
Canada follows this tradition. She stresses the use of play in subject learning in
primary school. However, with Thatcher's policy in England and after the
Education Act of 1988, formal academic learning in English and mathematics,
and testing of the formal levels became favored. Informal education was
reduced. But the results now show that English and mathematics scores have
been lowered, when the school stressed more formalized teaching in these
school subjects (Pollard, 1994; Galton, 1995). In addition social problems and
bullying in school increased. England is an interesting example as to what
happens if play and playful learning in school is reduced.
Austria went the opposite way. The formal, academic Austrian school for
the 6-year-olds have shown poor results. The Vienna play program in school
(Hartmann, 1988) led to a revision of their National Curriculum in 1987, which
was the most "playful" curriculum in Europe at that time. (Now Norway's new
National Curriculum of 1997 has taken over this position). Hartmann's research
is a good argument for why the Norwegian curriculum also may give good
results in creativity, social collaboration, and that they possibly will reach the

same level in their academic study as the non-play classes before 1997 in Norway! However, again, there is no Norwegian research yet to support this.

In Portugal, convincing research by Pessanha (1995) showed that the dropout percent in mathematics and language was greatly reduced after introducing her play program in school.

Denmark has for 27 years had the 6-year-olds in separate classes at school with their preschool teachers. During the eighties Denmark tried to bridge the gap between this preschool class and grade 1 and 2 at school. Preschool teachers and primary school teachers for the 7- and 8-year-olds built a team to create a common teaching program for the 6-8 year-olds for 20-50% of the total time in school. Workshop pedagogy, and a combination of practical and theoretical/academic subjects were preferred. Much structured constructive play and dramatic role play were used as working methods within a common theme, which lasted for 1-3 months. This strategy has been inspiration for the Norwegian reform. Evaluation showed that these children just like the Austrian were more creative, better in cooperation, wrote more interesting stories, liked school and reached the same level in academic studies as the control group.

From Arizona in the United States the "play and early literacy" movement is central (Christie, 1991; 1995; 1999; Christie & Roskos, 2001). Language programs based on role play in language stimulating play corners resulted in positive effects on academic performance in language in preschool/kindergarten, and also in children in grades 1 and 2.

So, does play pay?

I think the Norwegian "play revolution" for 6-10-year-olds in schools have possibilities to become a success, on one condition: All teachers, including preschool teachers for these children must have a thorough training in play pedagogy. Many Norwegian textbooks for students are attempts to provide this (Trageton, 1997; Eik, 1997; Vedeler, 1999; Kibsgaard, 1999; Lillemyr, 1999). The National Curriculum declares that the Norwegian school shall foster the meaning-seeking, creative, hardworking, all-around educated, cooperative, environmental-conscious human being. All these parts should together promote an integrated human being. In my opinion, the most integrated human being is Homo Ludens. In relation to the play theories mentioned earlier: Play is culture creation and a necessary fundament for learning in lower primary school.

REFERENCES

Bae, B. (1988). Voksnes definisjonsmakt og barns selvopplevelse. *Norsk Pedagogisk Tidsskrift* nr 4 p 212-227.

Bateson, G. (1955), The message 'This is play'. In B. Schaffner (Ed.), *Group processes: Transactions of the second conference* (145-242). New York: Josiah Macy, Jr. Foundation.

Bauer, G. G. (1996). *Homo ludens, der spielende mensch I, II, III, IV, V.* Salzburg: Katzbichler.

Bjørgen, I. (1992). Det amputerte og det fullstendige læringsbegrep. (The amputated and complete learning concept). *Norsk pedagogisk tidsskrift*, 1.

Brown, N. & Precious, M. (1969). *The integrated day in the primary school.* London: Ward Lock Educational.

Bruner, J. (1972). *Play, its role in development and evolution.* Harmonsworth. England: Penguin.

Christie, J. & Roskos, K. (2001). *Research on play and literacy: A critical review.* ICCP World play conference, Erfurt, Germany.

Christie, J. (1992). *Teacher education: A key element in improving school play.* ICCP/TASP Conference, Paris.

Christie, J. (1995). Linking literacy and play. *ICCP/TASP Conference,* Salzburg.

Christie, J. (1999). *Play in a literacy-enriched play center in same-age and multi-age grouping arrangements.* TASP Conference. Santa Fe, New Mexico.

Comenius, J. A. (1632). *Die Mutterschule.* (Norwegian translation. 1966).

Dewey, J. (1900). *The school and society.* Chicago: University of Chicago Press.

Eik, L. T. (1997). *Lekende læring og lærende lek* (Playful learning). Oslo: Pedlex.

Enz, B. & Christie, J. (1997). *The impact of teacher play interaction styles on children's play and emergent literacy activities. ICCP Conference,* Portugal.

Eriksson, E. H. (1941). *Barndommen og samfunnet.* (Norwegian translation, 1968.) Oslo: Gyldendal.

Freud, S. (1920). *Beyond the pleasure principle.* In J. Strachey (Ed.), The standard edition of the complete psychological works of Sigmund Freud (1956). London: Hogarth Press.

Frøbel, F. (1826). *Menschenerziehung.* 3nd edition. Düsseldorf: Hoffmann.

Gadamer, G. (1965). *Wahrheit und methode.* JBC Tübringen: Mohr.

Gardner, H. (1983). Frames of mind: The theory of multiple intelligence. *Harvard Educational Review, 57*(2) 187-193.

Garvey, C. (1979). *Lek.* (Play). Oslo: *Universitetsforlaget.*

Goodman, M. E. (1974). Barnkulturen. (The child culture). Stockholm: Wahlstrøm och Widstrand.

Groos, K. (1899). *Die Spiele der Menschen.* Jena: G. Fischer.

Hall, & Abbott, (1991). *Play in the Primary Curriculum.* London: Hodder & Stoughton.

Hartmann, W. (1988). *Spiel und elementares Lernen.* Vienna: Österreichischer Bundesverlag.

Haug, P. (1991). *6-åringane – barnehage eller skule?* (Six year olds – preschool or school?). Oslo: Samlaget.

Haug, P. (1996). *Barnehage på skule.* (Preschool at school). Trondheim: NOSEB.

Heggstad, K., Knudsen, B. & Trageton, A. (1994). *Fokus på lek.* (Focus at play). Bergen: Fagbokforlaget.

Hohmann, M., Banet, B. & Weikart, D. P. (1979). *Young Children in Action.* Ypsilanti. Michigan: High Scope Press.

Huizinga, J. (1938/1993). *Homo ludens.* København, Norway: Gyldendal.

Johnson, J., Christie, J. F. & Yawkey, T. D. (1999). *Play and early childhood development.* New York: Longman.

Kibsgaard, S. (Ed.). (1999). *Mens leken er god* (While play is good). Oslo: Tano.

Leontiev, A. N. (1977). *Problemer i det psykiskes udvikling. III.* København: Rhodos.

Lidén, H. (1994). *Mellom skole og fritid.* (Between school and leisure time). Oslo: Universitetsforlaget.

Lillemyr, O. F. (1999). *Lek-opplevelse-læring* (Play-experience-learning). Oslo: Tano.

Moyles, J. (1989). *Just playing? The role and status of play.* Philadelphia, PA: Open University Press.

Moyles, J. (1994). *The excellence of play.* Philadelphia, PA: Open University Press.

National Curriculum for Primary/secondary Schools in Norway. (1997). Læreplanverket for den 10 -årige grunnskolen (L 97).

Olofsson, B. K. (1987). *Lek för livet.* (Play for life). Stockholm: HLS förlag.

Opie, I. & Opie, P. (1969). *Children's games in streets and playgrounds.* London: Oxford Clarendon Press.

Otto, K. H. (1990). Zur spesifik der bezhiehungen zwischen kindern und erwachsenen im spiel. *ICCP.* Andreasberg, Germany.

Papert, S. (1983). *Dialog med datamaskinen.* Oslo: Cappelen.

Pessanha, A. (1995). *Comparative study of play. ICCP/TASP Conference* Salzburg.

Piaget, J. (1962). *Play, dreams and imitation in childhood.* New York: Norton.

Pollard, R. (1994). An analysis of the perceptions of preservice teaches toward technology and its use in the classroom. *Journal of Instructional Psychology, 21*(2), 131.

Rasmussen, B. (1990). *Å spille-eller late som om... Forståelsen for dramatisk spill i det 20. århundre.* (To play- or as if). Trondheim, Norway: Universitetet.

Retter, H. (1983). Spielen, arbeiten, lernen mit bildungsmitteln im erstunterricht. In: *Spiel-mittel, 3.* Jg., Heft 5, S. 7-11.

Rodari, G. (1991). *Fantasiens gramatik.* Göteborg: Korpen.

Schiller, F. (1793). *Briefe über die ästhetische erziehung des menschen.* Stuttgart, Germany: Reclam.

Spencer, H. (1855). *The principles of psychology.* London: Longman.

Sutton-Smith, B. (1971). Piaget on play: A critique. In R. E. Herron & B. Sutton-Smith (Eds.), *Child's play* (pp.326-336, 340-345). New York: John Wiley & Sons Inc.

Sutton-Smith, B. (1990). *Dilemmas in adult play with children.* ICCP Conference. Germany.

Titman, W. (1994). *Special People, Special Places.* WWF/Learning Through Landscapes.

Trageton, A. (1994a). *Leik med materiale 1-7 år.* (Play with materials). Bergen: Fagbokforlaget.

Trageton, A. (1994b). Workshop pedagogy – from concrete to abstract. *The Reading Teacher, 47* 4:350.

Trageton, A. (1995). *Role Play and drama 6-10 years.* ICCP/TASP Conference. Salzburg. Austria.

Trageton, A. (1997). *Leik i småskolen.* (Play in lower primary school). Bergen: Fagbokforlaget.

Trageton, A. (1999). *Play in lower primary school in Norway. TASP* Conference. Santa Fe, New Mexico.

Vedeler, L. (1999*). Pedagogisk bruk av lek* (Play pedagogy). Oslo: Universitetsforlaget.

Vygotsky, L. (1933). Play and its role in the mental development of the child. In J. Bruner, a. Jolly, & K. Sylva (Eds.). *Play: Its role in development and evolution (pp. 537-554).* Harmondsworth, England: Penguin.

Wassermann, S. (2000). *Serious players in the primary classroom:* Empowering children through active learning experiences. New York: Teachers College Press.

Wittgenstein, L. (1932). *Philosophische untersuchungen* (Danish translation, 1971).

Wood, D., McMahon, L. & Cranstoun, Y. (1980). *Working with under fives.* Ypsilanti, MI: High/Scope.

Chapter 9

The Mill Lake Hospital: Play in a Middle Grade Classroom

Lynn Romeo and Kristie L. Andrews

BACKGROUND

With the recent push for higher academic standards, prescribed curriculum, and additional standardized testing, the majority of teachers are very reluctant to integrate play into elementary classroom instruction (Manning, 1998; Hall, 2000; Roskos, 2001). Many educators believe that the positive aspects of the role of play in enhancing critical thinking and reflection frequently can not be measured by standardized and other paper and pencil tests (Manning, 1998). However, developmentally, children in both the primary and middle grades thrive on various types of play (Johnson, 1998; Manning, 1998). Although the modes of play that middle grade students enjoy tends to differ from younger children, "there continues to be a powerful need for play in late childhood" (Manning, 1998, p. 156).

Even children in the middle grades continue to enjoy pretend play and various sophisticated games as they socially interact and collaborate with other children. They create plays, stories, and puppet shows, utilize miniature figures, play school and house, and employ computers and other technological equipment (Davidson, 1998; Johnson, 1998; Manning, 1998). These interactions require planning, reflection, and critical reasoning which are all necessary for both academic and life success (Hall, 2000; Manning, 1998). In fact, the types of interactions and thinking skills promoted by play appear to be much more related to activities found in the workplace than many of the academic tasks in schools today (Hall, 2000). Games with rules can be effectively used in classrooms to assist children in developing their abilities to resolve conflicts and interact in a socially appropriate manner (De Vries, 1998). The use of play for older students at the secondary level has been also advocated to build collaboration skills (Fredericksen, 1999).

Unfortunately, a lot of instruction in school focuses on independent work and activities that do not provide opportunities for social collaboration and peer inquiry (Hall, 2000). Additionally, isolated instruction, especially the study of factual material from content texts and drill activities, makes it difficult for students, especially at-risk learners, to link their background information to the concepts being presented. Further, their prior knowledge may be lacking or incorrect (Finley, 1991), thereby impeding comprehension resulting in the formation of incorrect inferences (Pressley, 1998).

Scientists, in their daily work, utilize an array of methodologies that include observation, experiments, dissection, and collaborative inquiry (Wolfe, Cummins, & Myers, 1998). However, school science instruction, which often consists of the use of a traditional textbook that contains a lot of facts, frequently is viewed by the students as boring, and does not allow for higher level thinking and understanding (Roth, 1989; Dixon-Kraus, 1996; Beck & McKeown, 1988). The science texts vary greatly in their content presentation, vocabulary, and the types of questions posed (Meyer, 1991). In addition, generally one text is utilized and it does not provide the varied reading levels and competencies of the students in the class (Morrow, Pressley, Smith & Smith, 1997; Stahl, Hynd, Glynn & Carr, 1996). Frequently, the textbooks are often too difficult for many of the students to read (Holmes & Ammon, 1985; Stewart, 1994).

Murden & Gillespie (1997) modeled their research after two similar, earlier studies involving middle and high school students and teachers. Both groups were interviewed about content teaching and learning. The results indicated that teachers generally did rely on one textbook for delivering the content information mandated by the school's curriculum. However, students reported that they frequently did not read the textbooks prior to class and conjectured that the teachers were attempting to teach them only the pertinent information that they would need to pass a test.

Although it is important for students to learn content concepts, the teaching should include varied activities to allow for schema building as well as critical thinking, inquiry, and analysis (Morrow, Pressley, Smith & Smith, 1997; Stahl, Hynd, Glynn & Carr, 1996; Roth, 1989). In addition, there is an increased need for more social interaction and collaboration as learners are approaching middle school age. During this period of their development, students are striving to determine their place in the environment as well as how well their peers will accept them (Preisser, Anders, & Gilder, 1990). Thus, opportunities for social interaction and collaboration have been found to enhance motivation and engagement for academic tasks and peer learning (Turner, 1995, 1997).

A longitudinal study by Simpson and Oliver (1990) indicated that in grades six through ten, students' attitudes toward science declined yearly. When types of groups were analyzed, the middle group of students had the poorest attitudes. It also appears that "play is a powerful and underutilized vehicle in elementary science education" (Wolfe, Cummins, & Myers, 1998, p. 69) that can promote science achievement and critical thinking skills (Wolfe, Cummins, & Myers, 1998).

Literacy Play Centers, sometimes referred to as dramatic play areas, have been employed in classrooms of young children for many years. When the preschool environment is enhanced by including props that focus on reading and writing, young children's voluntary use of literacy activities and materials have significantly increased (Morrow & Rand, 1991; Neuman & Roskos, 1991, 1993; Vukelich, 1991; Young & Romeo, 1998; Einarsdottir, 2000). Children enjoy the experiences as they verbally interact and make attempts to read and write (Hall, 2000). Typical preschool centers, such as a kitchens, writing areas, or block areas can easily be transformed into restaurants, banks, aquariums, post offices, libraries, zoos, doctor or veterinarian offices, and construction sites. Props can include writing utensils, books, magazines, pads, letters, puppets models, pictures, signs, and charts. Children learn about literacy because it is integrated into an authentic context that is a familiar component of their environment (Pellegrini & Galda, 1993; Neuman & Roskos, 1997; Roskos, 1995; Walker, Allen & Glines, 1997; Labbo, 1998).

Roskos (2000) indicated that pretend play may actually foster the mental processes needed to understand print. Young children are able to demonstrate their increasing knowledge of print related concepts in a low-risk environment (Roskos, 2000). However, the amount of time that children spend in the Literacy Play Center seems to affect the nature of both the literacy and the dramatic play experiences.

Some researchers are beginning to use Literacy Play Centers with older students (Stone & Christie, 1996; Jarrett, 1997; Romeo & Young, 1999) and in multi-age classrooms (Christie & Stone, 1999). In one study (Romeo & Young, 1999), fifth grade students expanded their knowledge base about explorers in a center built to resemble a time machine. The Literacy Play Center supplemented the regular social studies program that consisted of a traditional textbook and direct teacher instruction. It was used for peer and independent practice in a context of play.

Props in the fifth grade Literacy Play Center included a computer with internet access, goggles, a compass, a protractor kit, a globe, games, books, journals, maps, puzzles, and travel brochures. It also contained lapboards, pillows, writing tools, a bulletin board that had a timeline, stickers, and individual folders for each student. The exterior sidewalls were bookshelves and the front entrance was made of thin plywood. Netting was placed from the ceiling to enhance the look of a time machine.

Analysis of the collected data which included interview transcripts, video tapes, student artifacts, and observational data revealed that the students responded very favorably to the physical environment of the Literacy Play Center which was small and self-contained due to the placement of the center's walls. It appears that classrooms that have been partitioned into smaller, more intimate spaces with engaging materials and comfortable furniture can engage students and promote thinking and learning (Morrow, 1997; Romeo & Young, 1999).

The fifth grade students also thrived with the opportunity to socially interact with their peers. During the interviews, students were very vocal about how they benefited from the opportunities to collaborate with others while studying about explorers. Finally, there was a high level of student motivation and interest for the activities in the Literacy Play Center. The students were actively engaged on tasks during the time they spent in the center (Romeo & Young, 1999).

In another study which involved randomly assigned second grade classrooms, Morrow (1992; Morrow & Sharkey, 1993) included literacy centers to promote literacy acquisition. In addition to books and writing utensils, the centers included felt boards and the use of drama and props to retell stories. The control groups completed workbook exercises and participated in a traditional basal reading series. The students in the literacy/play-enriched classrooms were more motivated and evidenced much more social interaction and collaboration.

During prior observations and interviews with older students, some of the students felt that the name Literacy Play Center was babyish and should only be used in classrooms of young children. This finding led us to create a new name for this type of integrated experience, the Literacy Exploration Lab. We wanted students to be able to use play and literacy to explore science or social studies content information in an engaging, small, enclosed environment.

METHODOLOGY

The purpose of this qualitative study was to investigate students' perceptions about their interactions while using a Literacy Exploration Lab that was created to link science content with play and literacy. We also wanted to find out if the lab activities had any impact on the students' factual science knowledge.

The following questions framed this study.

What kind of engagement was evident?

What type of activities did the students prefer?

Third grade students from one class in a suburban elementary school (grades kindergarten through three) in central New Jersey were the subjects for our study. There were twelve Caucasian students and one Asian-American student in the class, six girls and seven boys.

Four graduate pre-service students enrolled in a literacy course collaborated with the elementary students and classroom teacher to determine what type of Literacy Exploration Lab would be constructed. The choice, however, was limited to the theme that would be studied in science, which was support systems. The Literacy Exploration Lab was used as a supplemental activity to the science program used in the class, which consisted of whole group discussions, as well as lessons and experiments from the Scholastic Hands-on Science Program.

The classroom teacher and researcher was a second year teacher. Her background knowledge about linking play, literacy, and content learning was gleaned from extensive reading, the building of a Literacy Play Center for preschoolers during her enrollment in a graduate literacy class, and developing a zoo center in her second grade classroom the previous year. The other researcher was a university professor who taught the graduate course.

It was collaboratively decided to create The Mill Lake Hospital, which focused on the study of the human body and support systems. The teacher, graduate students, and elementary students also discussed possible props and materials for the center. The graduate students received instruction in their graduate course regarding how to build an engaging, self-contained, literacy/play enriched environment where the third graders could learn more about science concepts. The development of the center was grounded in the work of Morrow and Rand (1991), Christie et al. (1988), and Neuman and Roskos (1991, 1992, 1993).

It was also suggested to the graduate students that they specifically develop activities that focused on the following words: *create* (a meal that focuses on the Food Pyramid), *imagine* (that you grow up to have a very active job. What might it be?), *pretend* (you are a doctor treating someone with a broken bone and need to decide what kinds of treatment should be used), *design* (a slogan), or *write* (about a time when you were sick).

After the graduate students constructed The Mill Lake Hospital, they met with the teacher and students to introduce the lab and explain the props, activities, and materials. The class was given a Monmouth University diploma, which enabled them to practice medicine at The Mill Lake Hospital.

The exterior sidewalls were made from plywood that had been fastened together. The walls were painted white and the name of the hospital was written on the outside walls. Props included a gigantic felt board, activity binders, games, books, magazines, journals, a computer with internet access, a doctor's kit, hospital accessory kit, a crutch, and x-rays. It also contained a small mattress, pillows, writing tools, crates, stickers, and signs. The felt board was used to assemble and label a life-sized skeleton. A basket contained cards with ideas for journal writing. A Nutrition Box contained different empty food boxes so that students could determine the nutritional value. A poster with a picture of the brain that included various parts and functions was called a brain map. In addition, many other teacher made games, such as Bingo, Jeopardy, Fractured Phrases, Broken Bones, and Memory Inside the Body were provided. *Fun Facts* relating to interesting trivia about the human body were posted throughout the lab. Models of the tongue, brain, and bones were also provided and displayed.

The Mill Lake Hospital was utilized in the classroom for twelve weeks. The teacher placed the students in groups of four or five, based on literacy ability. The groups rotated between the Literacy Exploration Lab and three other literacy centers (the library, word building center, and writing center) that were constructed by the classroom teacher. Generally, every student was able to work in the lab thirty-five minutes, once per week.

Initially, the students were given complete choice of the activities they could do in the lab. The teacher and other observers noticed, however, that the students tended to either *play* doctor or worked on the same games repeatedly. After the researchers reflected and discussed what was occurring, it was collaboratively decided to set some parameters since there was no adult present that could constantly facilitate the learning experiences. For the remainder of the sessions, the teacher chose four activities weekly and instructed the students that they must select two of the four weekly. At the end of each session, students were given time to write their thoughts and reflections regarding what they learned. Journals were provided for this metacognitive, self-regulatory activity.

The methods employed in the study were observations, a questionnaire, interviews, and analysis of artifacts from the center. A science test (pre and post) that was constructed by the graduate students and the classroom teacher was administered to all of the students.

The researchers and a graduate student observed the classroom weekly and took detailed field notes during each visit. Each observation lasted approximately one-half hour. The field notes were transcribed after each visitation. We viewed the students' mannerisms and verbal interactions. We wrote down the students' conversations and described what each student or small groups of students did while they participated in the Literacy Exploration Lab. In addition, we videotaped the students while interacting in the Literacy Exploration Lab two times during the twelve-week period. When the researchers viewed the videotapes, notes were taken. We looked for evidence of engagement, any interaction among students, and the types of activities that were chosen by the students. These notes were compared to the field notes from the observations. Our field notes were then expanded and refined to include all additional information. We also reviewed all of the journal entries.

At the end of the study, the students were administered a questionnaire. The questionnaire contained a five point rating scale (pictures) as well as several open-ended questions about the strengths and weaknesses of the lab. The students were asked to rate how they felt about the types of activities, and their overall feelings regarding having the Literacy Exploration Lab in the classroom. The mean and range of scores for each of the questions were analyzed. The open-ended questions were coded and compared with the information gleaned from the observational data.

Semi-structured interviews were also conducted. Each student was individually presented with the information from his/her questionnaire and asked to elaborate on the survey results to confirm or negate the information gleaned from the questionnaire. In addition, several questions were posed about their experience, what they completed while in the lab, and how it compared with other science instruction.

The classroom teacher kept a journal of her impressions of what occurred while the students were participating in the lab.

We conducted qualitative analysis (Glazer & Strauss, 1967; Bogdan & Biklen, 1992) to analyze the observational data. Using our initial research questions, we read the transcripts of the observations line by line, made notes, and highlighted pertinent points. We then reread the data and searched for emerging patterns. Both researchers continued this analysis throughout the study, comparing and contrasting new data with the existing categories. Both researchers also viewed the videotapes. For data triangulation, the interview and questionnaire data was compared and contrasted with the observational data. The researchers discussed the emerging categories until agreement on the categories was reached. A graduate student, trained in coding procedures, served as an independent rater.

RESULTS

The mean scores on the five point rating scales from the questionnaire ranged from 3.53 to 4.7. The lowest means involved questions regarding how the students felt when they wrote about information in the lab (3.53) and a general question about their feeling regarding reading and writing (3.77). The highest means, (4.7 and 4.46 respectively) involved questions about the students' perceptions regarding whether or not they would like to have another lab in their classroom and a general question concerning how they felt about having the Mill Lake Hospital in their classroom.

The classroom teacher and graduate students who built the lab constructed the science pretest/posttest that consisted of fifteen factual questions. The mean score on the pretest was 4.4 and the mean score on the posttest was 10.6. The results of a t-Test revealed that the posttest scores were significantly superior to the pretest scores (p. < .01).

Four categories emerged from the qualitative data: Engagement, Self-regulation, Collaboration, and Types of Activities.

Engagement

There was a high level of student engagement evident during both the observations and the video taped sessions. Students were on task when working alone, in pairs, or groups. Often, students continued to be engaged after they were informed that they had run out of the time allotted for student use of the Literacy Exploration Lab. Neither the teacher nor the classroom aide were required to intervene due to off-task behavior or any discipline problems. This finding was very similar to one of the categories that emerged in an earlier study (Romeo and Young, 1999) that investigated fifth graders use of a center to study explorers. The engagement noted in the current study was the most evident during participation in games and least evident during text reading.

> The boys continued to work on the game until it was too late to write in their journals.
> All three students forgot to take off their nametags and had to come back later to return them.
> At the felt board, *C* is saying, "No, it doesn't go there." *M* said, "Oh, yeah, we need the spine." *F* stands behind *C* and is leaning over them.
> *G* was serious about this project. She was engrossed throughout the lab time and appeared very proud of her work.

Self-Regulation

The students were very self-regulated during the time they spent in the lab. Students actively engaged their background knowledge and applied it to what they did in the Literacy Exploration Lab. For example, students reflected on

commercials they had viewed about milk and used this knowledge to create slogans. One student labeled two bones on the felt board and remarked that she had
previously broken them. Another boy shared information that he had seen on an
episode of the Magic School Bus Show. There was also much evidence of self-
monitoring. The students were very actively in charge of their learning and were
able to manage themselves while working in the lab. Students self-checked their
work after completing diagrams and models.

> When he finished writing, he appeared to re-read and think about
> what he had written. Then he said, "Period," and added the punctuation
> mark.
> *B* called, "*L*, you have to test me on this," and handed the diagram to
> *L* while she began labeling the bones. As she labeled each bone, *B* said the
> name aloud.
> *L* checked her name tag (pediatrician) and announced, "I help chil-
> dren." *B* said, "That's what my aunt is."
> The students wrote about what they learned in their journals after re-
> flecting about their experiences.
> I learned all different kinds of foods and what you should eat.
> I learned that the heart pumps blood.
> I learned how to make a slogan. It is really easy. I made 4 slogans. I
> made one on orange juice, broccoli, carrots, and bananas.
> I learned that if you stay in a spot too long, you'll get stiff.

Collaboration

The social interaction that took place in the Literacy Exploration Lab involved working collaboratively to assemble models, write slogans, use the felt
board, play games, and share information. There was also a great deal of evidence of student coaching and peer scaffolding. This involved helping one another when one member was experiencing difficulty completing a task or locat-
ing an answer. Frequently, one student would pronounce a difficult word for
another student. Sometimes, the coaching took the form of one student posing a
question to others or reminding another student of what choices they were given
for the week. Students related the information to their prior world knowledge as
well as what they were learning in the classroom.

> *L* wrote 'broccoli is green …' After thinking for a minute, he added,
> 'greenest vegetable in town.' When *L* seemed a little stuck, *C* told him,
> "Write what will make them buy it." *M* further explained, "Like 'Got
> Milk' is 'Got Milk.'
> While at the eye chart, *T* and *D* were trying to read a hidden message.
> *T* said, "If you look at the words closely, that's what it says." *D* replied,
> "I'll tell you what it says and then read the message."
> The boy continues to work on the felt board and the ophthalmologist
> tells him that he is doing good. "This is an arm," *C* added, "I got a bone."
> *D* told him that they are all bones. *C* accidentally bumped the board and
> the bones were knocked off. *D* demonstrated how to put them back up. *D*

said, "This is what you do."

Types of Activities

It was clearly evident from all data sources that the students preferred the activities that involved games and manipulatives. An example of one is the Fractured Phrases game. This activity required the students to make up sentences and stories using words (300 were provided) that were on magnetic strips. Many of the students chose and spoke about liking the felt board, which required the students to assemble a life-size skeleton. They also chose games like Jeopardy, Operation, and Bingo or models, such as the brain, tongue, and bones. In addition, the students liked creating and writing slogans. An interest in role playing was also apparent. The students frequently wore signs such as Pediatrician, Surgeon, Radiologist, and Orthopedist.

During the interviews, some students spoke about liking the books and indicated that they found them to be interesting. They said that they learned new information from the readings. This was not apparent during the observations, though. There was no evidence in any of the observations of any sustained text reading. Students did look at some of the books, but quickly put them down in favor of a game or manipulative.

> The radiologist looked over some x-rays and asked himself what is this x-ray of.
> 'I liked the felt board, Jeopardy, and Bingo.'
> *J* and *L* went to the felt board, knelt on the floor, and emptied the envelope of the felt bones. . . . They swiftly began to assemble the skeleton.
> *N* suggested playing Jeopardy. *M* was very quick to push the curtain back and ask, "Did you say Jeopardy?" A started humming the theme music that plays during the show while he and *L* put away the bones. *M* announced, "I'm Alex."

CONCLUSIONS

This study explored the use of a Literacy Exploration Lab to supplement a third grade class' science unit on support systems. We were interested whether the students would be actively engaged in using the props, books, and games that were present in the lab and what specific types of activities they would prefer to use. Factual science knowledge was assessed using a pre and post-test. The students scored significantly higher on the post-test than they did on the pretest, which was administered prior to the opening of the lab.

Four themes emerged from the data that was collected through extensive observations, questionnaires, and semi-structured student interviews. First, the level of student engagement was very high. The students actively explored the information that was provided in the lab and exhibited on-task behavior. The students began to work as soon as they entered the center and continued until they were informed that it was time to leave. This finding was similar to previous research (Romeo & Young, 1999) in a fifth grade center that focused on social studies content. A high level of student engagement is an important factor for teachers to focus on since it promotes motivation for reading and thinking (Baker, Afflerbach, & Reinking, 1996; Baker & Wigfield, 1999).

Secondly, student self-regulation was very apparent. The students actively took charge of initiating and monitoring their own learning and drew from their background knowledge. This may have been enhanced by the inclusion of the journal in which the students were required to think and write about what they learned in the lab. This seemed to foster metacognitive skills. In addition, the classroom teacher required these students to reflect daily about their total class-room learning as well as share that information with their parents nightly. Thus, it is possible that the teacher's encouragement of self-regulation promoted the students' taking charge of their learning in the lab. Further research should com-pare Literacy Exploration Labs in a classroom that promotes self-regulation with a classroom where instruction does not focus on multiple opportunities for re-flective thinking and inquiry.

The third theme was student collaboration, which was clearly evident among all of the data sources. Students coached each other and worked in groups to solve problems, construct models, write slogans, and role play.

This may have been promoted due to the integration of play "because it of-fers a tension-free environment in which students can get to know one another and learn to cooperate to achieve a common goal" (Fredericksen, 1999, p. 116). This collaboration and community building was also found in an earlier study of fifth graders working together to learn about explorers (Romeo & Young, 1999). Other successful integrated content learning approaches have involved student collaboration and peer inquiry (Brown, 1992; Guthrie, 1996).

The students in this study preferred the games, manipulative activities, and the creative writing of slogans. In a study that compared engaging first grade classrooms to non-engaging first grade classrooms (Bogner, Raphael, & Pressley, 2001) games were used much more frequently in the engaging class-rooms. The students in this study chose these materials more regularly than books. At first glance, one might think that they did not enjoy text reading. However, it should be noted that the games all involved problem solving and reading. In addition, this classroom had an elaborate reading corner that students used for blocks of time. It appears that they viewed the Literacy Exploration Lab as a time for cooperative play. This focus on play did provide all students with the natural opportunities to practice what they were learning about science.

This study was limited to a single, small class of third grade students. Addi-tional research is needed that focuses on comparing the achievement and en-gagement of multiple classes of students, some of whom have their content subject supplemented by a Literacy Exploration Lab and others who receive traditional, textbook instruction. Secondly, although the students' scored signifi-cantly higher on the posttest of factual science knowledge, we were unable to separate the knowledge gained in the center from information gleaned from the Scholastic program and class discussions.

In addition, it was difficult to determine the effect of the physical environment on this group of students. This particular class has an elaborate reading corner with physical boundaries and comfortable pillows as well as a well-equipped writing center, and a hands-on word study center. In a prior study (Romeo & Young, 1999), fifth grade students spoke constantly about the comfort and coziness of the center. That particular classroom had no other centers or engaging areas. Since there are virtually no studies that have used Literacy Play Centers or Literacy Exploration Labs with older students, additional research is needed to explore the issue of the physical environment.

Another limitation was the researchers' decision to limit the choice of activities that could be done weekly. Thus, the students were only permitted to choose among four activities rather than the array that was available. This was a difficult issue for us, one that we have still not resolved. On the one hand, we view choice as a necessary component for active engagement. However, we are also aware of the limited classroom time available for learning and felt that the students were spending too much time on pretend play that would not promote the critical thinking necessary for content acquisition. Since this did not occur in the fifth grade study (Romeo & Young, 1999), it is possible that third grade students might require more adult facilitation and scaffolding than the older students. Directly linking the dramatic play to the writing of short skits and plays might also be a way to enhance the dramatic play and connect it specifically to science and literacy.

Literacy Exploration Labs can foster student engagement and self-regulation, components that teachers need to actively promote in today's classrooms. They provide a naturalistic context for studying content material in a *play* setting and give multiple opportunities for teachers to create a range of instructional materials that are appropriate for the diverse students in a classroom. The myriad of opportunities for learning literacy and content "can better prepare students for reading in a wide range of situations, as expected in literate societies" (Baker, Afflerbach, & Reinking, 1996, p. *xvi*). The Literacy Exploration Lab's focus on collaborative, self-regulated thinking and learning is extremely important for fostering the type of skills necessary to be successful in future educational and life endeavors.

REFERENCES

Baker, L. & Winfield, A. (1999). Dimensions of children's motivation for reading and their relations to reading activity and reading achievement. *Reading Research Quarterly, 34*, 4, 452-477.

Baker, L., Afflerbach, P. & Reinking, D. (1996). Developing engaged readers in school and home communities: An overview. In L. Baker, P. Afflerbach, & D. Reinking (Eds.), *Developing engaged readers in school and home communities* (pp. xiii-xxvii). Mahwah, NJ: Erlbaum.

Beck, I. L. & McKeown, M. (1988). Toward meaningful accounts in history texts for young learners. *Educational Researcher, 17,* 31-39.

Bogdan, R.C. & Biklen, S.K. (1992). *Qualitative research for education: An introduction to theory and methods.* (2nd ed.) Boston, MA: Allyn & Bacon.

Bogner, K., Raphael, L.M. & Pressley, M. (2001). How grade-1 teachers motivate literate activity by their students. *SSSR Journal: Scientific Studies of Reading,6,* 2, 135-166.

Brown, A. L. (1992). Design experiments: Theoretical and methodological challenges in creating complex interventions in classroom settings. *Journal of Learning Sciences. 2,* 141-178).

Christie, J. F. & Stone, S. J. (1999). Collaborative literacy activity in print-enriched play centers: Exploring the 'zone' in same-age and multi-age groupings. *Journal of Literacy Research, 31,* 2, 109-131.

Christie, J. F., Johnsen, E. P. & Peckover, R. B. (1988). The effects of play period duration on children's play patterns. *Journal of Research in Childhood Education, 3,* 123-131.

Davidson, J. I. (1998). Language and play: Natural partners. In D. P. Fromberg & D. Bergen (Eds.), *Play from birth to twelve and beyond: Contexts, perspectives, and meanings* (pp. 175-183). NY: Garland Publishing.

DeVries, R. (1998). Games with rules. In D. P. Fromberg & D. Bergen (Eds.), Play from birth to twelve and beyond: Contexts, perspectives, and meanings (pp. 409-415). NY: Garland Publishing.

Dixon-Krauss, L. (1996). *Vygotsky in the classroom: Mediated literacy instruction and assessment.* NY: Longman.

Einarsdottir, J. (2000). Incorporating literacy resources into the play curriculum of two Icelandic preschools. In K. Roskos & J. Christie (Eds.), *Play and literacy in early childhood: Research from multiple perspectives* (pp. 77-90). Mahwah, NJ: Erlbaum.

Finley, F. N. (1991). In C.M. Santa & D.E. Alvermann (Eds.), *Science learning: Processes and applications* (pp. 22-27). Newark, DE: International Reading Association.

Fredericksen, E. (1999). Playing through: Increasing literacy through interaction. *Journal of Adolescent and Adult Literacy, 43,* 2, 116-124.

Glaser, B. G. & Strauss, A.L. (1967). *The discovery of grounded theory: Strategies for qualitative research.* New York: Aldine.

Guthrie, J. T. (1996). Educational contexts for engagement in literacy. *The Reading Teacher, 49,* 432-445.

Hall, N. (2000). Literacy, play, and authentic experience. In K.A. Roskos & J.F. Christie (Eds.), *Play and literacy in early childhood: Research from multiple perspectives* (pp. 189-204). Mahwah, NJ: Erlbaum.

Holmes, B. C. & Ammon, R. I. (1985). Teaching content with trade books. *Childhood Education, 1,* 366-70.

Jarrett, O. (1997). Science and math through role-play centers in the elementary classroom. *Science Activities,* 13-19.

Johnson, J. E. (1998). Play development from ages four to eight. In D. P. Fromberg & D. Bergen (Eds.), *Play from birth to twelve and beyond: Contexts, perspectives, and meanings* (pp. 146-153). NY:Garland Publishing

Labbo, L. (1998). Social studies 'play' in kindergarten. *Social Studies and the Young Learner, 10*, 4, 18-20.

Manning, M. L. (1998). Play development from ages eight to twelve. In D. P. Fromberg & D. Bergen (Eds.), *Play from birth to twelve and beyond: Contexts, perspectives, and meanings* (pp. 154-161). NY: Garland Publishing.

Meyer, L. A. (1991). Are science textbooks considerate? In C.M. Santa & D.E. Alvermann (Eds.), *Science Learning: Processes and applications* (pp. 28-37). Newark, DE: International Reading Association.

Morrow, L. M. & Rand, M. (1991). Preparing the classroom environment to promote literacy during play. In J. Christie (Ed.), *Play and early literacy development* (pp. 141-163). Albany: SUNY Press.

Morrow, L. M. (1992). The impact of a literature-based program on literacy achievement, use of literature, and attitudes of children from minority backgrounds. *Reading Research Quarterly, 27*, 250-275.

Morrow, L. M. & Sharkey, E. A. (1993). Motivating independent reading and writing in the primary grades through social cooperative literacy experiences. *The Reading Teacher, 47*, 162-164.

Morrow, L., Pressley, M., Smith, J. K. & Smith, M. (1997). The effect of a literature based program integrated into literacy and science instruction with children from diverse backgrounds. *Reading Research Quarterly, 32*, 1, 54-76.

Murden, T. & Gillespie, C. S. (1997). The role of textbooks and reading in content area classrooms: What are teachers and students saying? In W. Linek, & E. G. Sturtevant (Eds.), *Exploring Literacy* (pp. 85-96). TX: College Reading Association.

Neuman, S. & Roskos, K. A. (1991). The influence of literacy enriched play centers on preschoolers' conceptions of the functions of print. In J. Christie (Ed.), *Play and early literacy development.* (pp.167-88). Albany, NY: SUNY Press.

Neuman, S. & Roskos, K. A. (1992). Literacy objects as cultural tools: Effects on children's literacy behaviors in play. *Reading Research Quarterly, 27*, 203-225.

Neuman, S. & Roskos, K. A. (1993). Access to print for children of poverty: Differential effects of adult mediation and literacy-enriched play settings on environmental and functional print tasks. *American Educational Research Journal, 30*, 95-122.

Neuman, S. & Roskos, K. A. (1997). Literacy knowledge in practice: Contexts of participation for young writers and readers. *Reading Research Quarterly, 32*, 10-33.

Pellegrini, A. D. & Galda, L. (1993). Ten years after: A reexamination of symbolic play and literacy research. *Reading Research Quarterly, 28*, 163-175.

Preisser, G., Anders, P.L. & Glider, P. (1990). Understanding middle school students. In G.G. Duffy, (Ed.), *Reading in the middle school,* (pp. 16-31). Newark, DE: International Reading Association.

Pressley, M. (1998). *Reading instruction that works: The case for balanced teaching.* (2nd ed.). New York: Guilford Press. (pp. 16-31).

Romeo, L. & Young, S. (1999). Using literacy play centers to engage middle grade students in content area learning. In J.R. Dugan, P.E. Linder, W.M. Linek & E.G. Sturtevant (Eds.), *Advancing the world of literacy: Moving into the 21st century.* Commerce, TX: College Reading Association.

Romeo, L. (1999). A recipe for promoting effective classroom literacy environments. *The Delta Kappa Gamma Bulletin, 65,* 4, 39-44.

Roskos, K. A. (2000). *Creating connections, building constructions: Language, literacy, and play in early childhood.* Reading Online. http://www.readingonline.org/articles/artindex.asp?.HREF=/articles/roskos/index.html.

Roskos, K. A. (1995). Creating places for play and print. In J.F.Christie, K.A. Roskos, B.J. Enz, C. Vukelich, & S.B. Neuman (Ed.), *Readings for linking literacy and play* (pp. 8-17). Newark, DE: International Reading Association.

Roth, K. J. (1989). Science education: It's not enough to do or relate. *American Educator, 13,* 4, 16-48.

Simpson, R. D. & Oliver, J. S. (1990). A summary of major influences on attitude toward and achievement in science among adolescent students. *Science Education, 74,* 1, 1-18.

Stahl, S. A., Hynd, C. R., Glynn, S. M. & Carr, M. (1996). Beyond reading to learn: Developing content and disciplinary knowledge through texts. In Baker, L, Afflerbach, P. & D. Reinking (Eds.), *Developing engaged readers in school and home communities* (pp. 139-164). Mahwah, NJ: Lawrence Erlbaum Associates.

Stewart, R. A. (1994). A causal connective look at the future of secondary content area literacy. *Contemporary Education, 65,* 90-94.

Stone, S. J. & Christie, J. F. (1996). Collaborative literacy learning during sociodramatic play in a multiage (K-2) primary classroom. *Journal of Research in Childhood Education, 10,* 2, 123-133.

Turner, J. C. (1995). The influence of classroom contexts on young children's motivation for literacy. *Reading Research Quarterly, 30,* 410-441.

Turner, J. C. (1997). Starting right: Strategies for engaging young literacy learners. In J.T. Guthrie, & A. Wigfield, (Eds.), *Reading engagement: Motivating readers through integrated instruction.* (pp. 183-204). Newark, DE: International Reading Association.

Vukelich, C. (1991). Materials and modeling: Promoting literacy during play. In J. Christie (Ed.), *Play and early literacy development* (pp. 215-246). Albany, NY: SUNY Press.

Walker, C. A., Allen, D. & Glines, D. (1997). Should we travel by plane, car, train, or bus? Teacher/child collaboration in developing a thematic literacy center. *The Reading Teacher, 50,* 524-527.

Wolfe, C. R., Cummins, R. H. & Myers, C. A. (1998). Dabbling, discovery, and dragonflies: Scientific inquiry and exploratory representational play. In D.

P. Fromberg & D. Bergen (Eds.), *Play from birth to twelve and beyond: Contexts, perspectives, and meanings* (pp. 68-76). NY: Garland Publishing.

Young, S. A. & Romeo, L. (1998). Promoting literacy in play centers: The importance of adult facilitation. *Paper presented at the annual meeting of the Association for the Study of Play*. St. Petersburg, Florida.

Chapter 10

Preschool Teachers Conceptions of Children's "Chaotic Play"

Annica Löfdahl

WHAT IS PLAY?

Definitions of play are full of nuances. What is play and what is not? Where does it come from and from where do we get all different conceptions? There are diverse play forms such as daydreams, jokes, festivals, competitions and challenges and several different play actors, children, men, women, clowns etc. all incorporating different play actions. In different places in different times, play has been perceived differently and it is difficult to define. For example, in western societies it is common to regard play as an activity associated with children and their development (Sutton-Smith, 1997).

Play can be elucidated with several different perspectives. According to the Swedish play researcher Olofsson (1987), play theories developed as an endeavor to explain why and how play arises as well as its purpose and development. Different researchers view play from different perspectives which makes theories complementary rather than replaceable, and different perspectives give answers to different questions, although Wood and Attfield (1996) regard this as if different perspectives have their grounding in different ideologies, and the ambiguity of play is due to the lack of agreement among ideologies. The pioneers (Fröbel, Montessori, Dewey, Steiner et al.) who pleaded for young children's education had different views on children's learning and the role of play in preschool. The ambiguity from former days is still present and contributes to the weakness and difficulties of defining and framing preschool pedagogy and, thereby, also play.

Categories and Definitions

Several play researchers (Bühler, Smilansky, Piaget, Rubin et al.) have defined and categorized play in a range of variations: role play, fantasy play, drama play, structured play, free play, construction play, rough and tumble play, war play and so on (Olofsson, 1987). We can imagine that Piaget's theories have been the inspiration for both practitioners and researchers by categorizing play as sensorimotor play, symbolic play and play with rules.

Play researchers also made up criteria for what is supposed to be play. Garvey (1990) lists different distinctive features of play: play is a positive experience for the player; no extrinsic purposes just intrinsic motives without objectives; play is spontaneous and players engage of free will; play is active engagement and play is related to not-play i.e. we can only talk about play in contrast to something else. Olofsson (1987) terms the social rules of play as "agreement, turntaking and reciprocity." Vygotskij (Vygotsky) lists the following criteria: Play is imaginative situations; subordinate to rules; social to its origin and mediated through language. Children learn to play in co-operation with others (Vygotskij, 1980). Play is looked upon as progression where different categories follow each other and is dependent on children's cognitive development as well as estimated in "more or less" play dependent on the degree of correspondence to the criteria of play.

Criticism

Sutton-Smith (1999) argues in his criticism that play previously has been regarded as worthless but nowadays play has greater value and is looked upon as the symbol of the modern freedom. This covers the fact that children often engage in play due to a wish to be accepted by playmates, rather than a particular wish of freedom in play. Rasmussen (1993) argues that the traditional play research has a romantic view of play and takes for granted that imaginative play is the most important development factor in children's life and, thereby, avoids research on the violent play with components of chaos. Categories and definitions are far from distinct or complete.

Preschool teachers often have to explain and defend their practices as "useful," which has led to a perspective of play as functional. It is taken for granted that play is related to learning, but thereby the relation is not reflected upon or problematized (Wood & Attfield, 1996). Previous research shows that the adult's role is ambiguous and play in particular is seen as ambiguous: preschool teachers are unsure about both their own role in play as well as the role of play in the curriculum (Lindqvist, 1995; Löfdahl, 1998).

One of the goals related to learning and development in the Swedish national curriculum for preschool (Lpfö-98, 1998) is to strive for every child to develop their curiosity and enjoyment as well as their ability to play and learn. The curriculum for preschool does not regulate how these goals are supposed to be achieved; it is a matter for the professional teachers working there. Thus, it is a presupposition that play as well as the other content of the curriculum is discussed, problematized and reflected upon by the teachers in order to fulfill the goals.

WHY "CHAOTIC PLAY" AND WHAT DOES IT MEAN?

What is meant by "chaotic play"? What discerns it from other play forms we usually talk about, like role play, fantasy play and so on? Critics argue against researchers who define play only in positive terms, since play is contradictory and paradoxical and play hovers between chaos and harmony. Play is not always good; it can be aggressive, scary, hateful and mean (as in bullying).

The Danish play researcher Rasmussen (1993) describes the bodily and sensual aspects of play. He suggests that play moves between two poles or extremes, between chaos and order. In "arranged play" everyone deeply concentrates and a calm atmosphere is pervasive. These are the play forms which are interpreted and conceptualized by modern play research as sensorimotor-play, role play, etc. The other play, according to Rasmussen, is moving towards the other pole, toward chaos. It is bodily play, the play with large movements, the sensual and bodily "violent" play which has not been given priority by play researchers' and thereby is unexplored compared to other play forms.

The adults in our society do not value play forms with elements of fighting and chaos because they destroy our ideal picture of play. Preschool teachers seem to have an ideal image of play and it can be difficult to let the chaotic play take place in pre-school. It is easier to interrupt and take control over children's play activity before chaos enters (Fisher & Madsen, 1984).

Play is divided into "good" and "bad" and the "bad play forms" are represented by children's chaotic play. In some preschool settings children are allowed to be "violent" in their play, in other settings they are not (Rasmussen, 1993). Male preschool teachers often maintain that they are less restrictive than women are, and more active together with children in the chaotic play (Birgerstam, 1997), and that they view the chaotic play as play and not just messy activity (Lumholdt, 1998). The Finnish brain researcher Bergström (1997) describes two sides of children's play in his book *Black and White Play*, and he looks upon play as children's possibilities, in which we have access to destruction and chaos as well as to order and structure. Thanks to play, brain systems can develop when chaos is allowed to flow, therefore children should have the opportunity to play and to let all sides of play flow. Bergström (1997) defines the significance of play and its content like "bringing chaos into what you are playing in and what you are playing with" (p. 89).

Even if the chaotic play is not the teachers' favorite play form, it seems to attract children. A Swedish study (Strander, 1997) looked at how young people remembered their preschool time. It demonstrates that what is remembered as funny was playing without permission and opportunities to joke and make trouble together with frisky, lively teachers.

Aim

The aim of this study is to illuminate teachers' conceptions of chaotic play. Deliberately choosing a situation like "chaotic play" and letting preschool teachers discuss this phenomenon should bring the ambiguity of play to the surface and make it visible. I believe it would not have been the same if we had a discussion about play forms like constructive play or fantasy play.

Interviews

My sample includes 12 teachers, different from each other regarding age, sex (two males) and education, two nursery teachers and ten preschool teachers. Their vocational experience varied between 2 to 30 years and they worked with toddlers (age 1-3), siblings group (age 3-6) and in mixed group's age 1-6 years. Six teachers belonged to the same work team in an ordinary preschool, two teachers worked in a special "outdoor education" preschool, and two teachers worked in a preschool with "play pedagogy" as their special profile. The two men came from different pre-schools.

Interviews were carried out in their respective settings; each interview lasted for 30-45 minutes and was transcribed afterwards.

To encourage the teachers to talk about the topic they where shown a 5 minute videotape of children engaged in chaotic play, plus my description of the topic. The interviews were unstructured, and the teachers were asked to tell me about their experiences in similar situations, their thoughts and feelings, to make comparisons between how they usually act and how they want to act. I also asked them to tell me how their colleagues used to act and their reactions to that.

RESULTS

I chose to consider the results from a dialectic perspective, from the relationship between theory and practice. A concrete example makes it possible to illuminate this relationship. Theory supplies knowledge acquired during education and the public view in our society as well as research findings including play theory, child development and learning. Practice supplies knowledge derived from experiences, often unreflected and unique to a particular child group (who can play with whom, etc.), but also more general, as to engage in children's play. Practical knowledge is the feeling teachers acquired from practical work with children, or what is often called tacit knowledge (Kihlström, 1995; Rolf, Ekstedt, & Barnett, 1993), such as knowing when activities are "topple over" and how to act.

Conceptions of Chaotic Play

In my analysis of the respondents' statements, I can see a pattern emerge in which four different conceptions of the chaotic play appear. The conceptions differ from each other as they represent different views; different lines of action and different focuses in what is often taken for granted. Quotations from interview extracts follow after each description to show the nuances in every conception.

A need

The chaotic play is physical practice, children need an occasion to exercise, and especially boys need space. Outdoor environments facilitate the possibilities. Teachers focus on children's physical development.

"They have a lot of surplus energy which they happily must get rid of"

"I think they need this ... yes, they got to have some outlet according to their physical need. I like to be outside and of course I can suggest to them, there is more space outside for them to play ball, run, climbing trees and ride bicycles. Because it is ... very crowded inside I suggest to them that it is better to do this outside, I have to say, I really do."

A danger

Chaotic play easily degenerates and often turns into trouble. This play is noisy and messy and children can hurt themselves when the play gets dangerous. It is not "real play"; it is just running around. It is important to be in control before something happens. Chaotic play must be moderated, with regard to other children's activities as well as the need for silence. Teachers focus on order and structure as seen from the following examples:

"Well if you are alone after snack time and you have a lot of children .... Then you don't want it to be messy."

"It's bothersome for your hearing, if someone screaming it's really painful. Of course I tell them to calm down, but it is for their own sake, it will be bothersome for them. They often collide with each other and. Yes, I have to say we all restrain this, we have replaced the tables to make a break in their running "roadway," but it does not work. They are running a lot and I don't like it, I'm scared something will happen. They can go out and play, but indoors ... I want another indoor play."

Supply Meaning

Children must be provided with an opportunity to maintain their chaotic play, but not just in any way. It is important that chaotic play has "meaning" and there are two different ways to supply meaning:

1. By showing how to do, to guide play, to play with rules where teachers can frame and restrict children's play activity as in the play "who is afraid of the big bad wolf?" It is important that this play has functions. Teachers' focus on directing as seen in the following examples:

"If they are thief and police and I can see that it is just running ... someone is sad and ... I arranged so someone got the role of police commissioner and he can give order and tasks [...]."

"No I don't want chaotic play to disappear, but I can feel it is just aimless running around and I think I can make it work if I organize ... I can make a track in the yard and they can run for hours. And they really like it," and,

2. To inspire with props, imagination, and drama and by arranging the setting for play, teachers can challenge children's minds through the chaotic play. It is important to give chaotic play a content, and now and then to participate. Teachers' focus to inspire and influence as seen in these examples:

"I use to participate, to joke and frighten with them and grapple with them, to make them keep going [...]."

" I can guide this play further so that it will become something else than just running. So they can imagine .... What are you playing when you are running, or ... how is a tiger or a cat running and then they start to run like a tiger. I feel as I conduct their play but that is what I want to do, I feel everyone is more involved ... if you just get them an idea, they can go on with it."

Perceive Meaning

It is just not "running with their legs," it is something else. Chaotic play is essentially meaningful for the children. It is the way they organize their chaos-play. Chaotic play is a challenge to children's minds when they have to find ways not to collide with their peers. Chaotic play is accepted for what it is and teachers like to participate. It is important to follow the children where they want to play and teachers' gain trust to their own capacity to control when it's needed. As seen in the two examples below, teachers focus on the children's own ability.

"Sometimes you can see that ... round and round and ... but it always has a main theme... always a story, they run because they are dinosaurs or something [...]."

"We have arranged our tables to make it harder for children to run, and I thought of that ... it is much more fun now, because they have to run zigzag to get through. I think it is to be together with your peers, and then you know what you are doing. They can see and recognize, it is easy, you can jump a bit and scream and then you are part of the play. There are no great demands upon children to be part of the chaotic play, and I am pleased to support those children who want to, but ... I don't break it up, I really don't!"

Comments on the Teachers Conceptions

The teachers all express more than one view, though I strive to find one dominant view for each person. One reason for the teachers having more than one view may be related to a frequent utterance, "it depends on how I feel." Some days teachers can look upon children's chaotic play as troublesome and

messy and they will not let them continue; the next day the same person can see the play as important and meaningful. One teacher expresses this ambiguity:
" ... it depends on my mood, if I am in a good mood I can see it is an excellent play, if I'm in a bad mood maybe I don't ... realize it is so very excellent as I believe it is [...]."

All teachers share the view of chaotic play as "a need," but it does not dominate for anyone. I mean this is a basic knowledge everyone acquired during his or her education. Children's physical development is one important subject during early childhood teacher education and everyone realizes the importance of motor training among children, but this is not viewed as play and they only mention this incidentally. I regard this view as basic from which the other three views appear as figures.

The views "a danger" and "supply meaning" are to be found among almost all teachers, whereas to "perceive meaning" is not. What does this mean? Is it just an expression of the ambiguity of play or is there some other explanation?

An interpretation from the perspective of theory and practice show that none of the predominant theories of play fit chaotic play; there is no tradition in Swedish preschools to see the chaotic as play and tradition in preschool settings is strong. Among those teachers who express "a danger" as a dominant view there is no problem to act like they do, to focus on order and structure. Thereby they have no need to change their way of thinking, to reflect on their actions. Theories about "teacher thinking" and "reflective practitioners" suggests that it is only when you discover boundaries or possibilities in teaching you also need to change your thinking (Alexandersson, 1995; Carlgren, 1995). Teachers, who express this view vaguely, also express a wish to act differently.

There is a relation between "a danger" and "supply meaning" through directing children's play. By engaging children in other activities rather than chaotic play, such as adult directed play activities, teachers can continue their control.

"Supply meaning" is to be found among all teachers, which is easy to see. This view is supported by theories from developmental psychology and the national curriculum, which encourage teachers to inspire children with materials and emphasize children's social and cognitive development. It is possible to see the teacher as the "practical applicator " who sees a mechanical relation between goal and means and allow theory to have a prescriptive function (Kveli, 1994).

There is also a relation between to "supply meaning" by inspiring children with imagination, and "perceive meaning." When teachers' influence chaotic play either with props or through participating they also trust children's own meaning-making in the chaotic play. This became visible during one interview when the teacher realized her ability to perceive meaning was dependent on whether children wore cloaks, like Batman, or not while they where running.

"I haven't thought of it before. . . . But when you say . . . well, I think children are more engaged when they . . . but of course, maybe they are as much engaged when they just run . . . but I can't see those pictures in front of me."

Not all teachers express the view "perceive meaning," but among those four teachers where this view is dominant it is obvious that they did reflect upon their actions. Instead of referring to a more traditional method, like moderating or interrupting the chaotic play, these teachers can see new possibilities. Chaotic play really is a complicated situation where teachers must consider different aspects at the same time to enable themselves to accept and allow it. This quotation elucidates the complexity:

> I've made it my mission to show that this is play, it is not just running. Because it is messy when 15 children are running and are very noisy, it's almost a public nuisance. If you're not in it yourself of course, then you can't hear it because the feeling of screaming is so important right then. Often I try to be part of their play, not directing it but just following along. If they are Batman, I can also be Batman for a while. But I don't have the strength like they have; I have to find a way out, like go to sleep or something. Then I'm out of play to a certain degree.

As a teacher you don't know where you are going or what will happen when you follow the children. It is not possible to have a method in advance, instead you must reflect during action. This presupposes teachers to trust their vocational competence and their ability to take control when it is needed, to have practical knowledge in combination with theoretical.

Comments Around Chaotic Play

What do teachers express when they discuss chaotic play? Some information, which is not visible in the conceptions, is nevertheless a part of the context, and is of great value when we try to understand chaotic play.

Chaotic play is not part of the planned curriculum, but teachers witness it occurring frequently every day. It is only indoors chaotic play that is perceived as chaotic. Those teacher working with "outdoor education," do not realize it could be problematic. Children have a lot of space outdoors and they can be as noisy as they want, they do not disturb anyone else.

On the contrary, teachers express many thoughts and feelings about chaotic play indoors. Too many children in too small spaces as well as age-mixed groups are perceived as obstacles. When different play forms are played at the same time, it is always chaotic play, which is not allowed, compared to other "calmer" play forms. One teacher argues that, "Chaotic play produces conflicts among children, and it is sometimes useful for them to learn how to solve a problem."

The debate preferring more male preschool teachers due to their ability to be more engaged in chaotic play and to be less restrictive does not correspond to my findings. Female preschool teachers also express enjoyment and pleasure while participating in chaotic play forms such as rough and tumble play.

CONSEQUENCES

We may wonder if all views are needed in preschool? Maybe it is a presupposition in a work team that one colleague sees chaotic play as a danger so he/she structures everyday life in pre-school, while another colleague engages fully to perceive meaning together with children in play. I believe this must be a continued discussion in every preschool setting.

What are the consequences regarding co-operation in a work team when, on the one hand, chaotic play is not allowed and on the other hand, it is allowed depending on the teacher's mood. Which view is most highly valued?

Teachers do not plan chaotic play, and it is hardly ever discussed. I suggest that this reinforces dominant attitudes, to see all forms of chaotic play as "bad."

By taking part in the results of this study, teachers may become aware of their own as well as their colleagues' views on chaotic play and thereby increase their understanding of each other's perception. I believe this will make it possible to influence and change attitudes towards play, and to reflect about and problematize all forms of children's play.

NOTE

This paper was presented at EECERAs 10th European Conference on the Quality of Early Childhood Education: Complexity, Diversity and Multiple Perspectives in Early Childhood Services. Institute of Education University of London August 29 - September 1, 2000.

REFERENCES

Alexandersson, M. (1995). *Profession och reflektion, Lärarprofessionalism - om professionella lärare.* Stockholm: Lärarförbundet. (Profession and reflection).

Bergström, M. (1997). *Svarta och vita lekar.* Stockholm: Wahlström & Widstrand (Black and white play).

Birgerstam, P. (1997). *Kvinnligt och manligt i förskolan.* Lund: Studentlitteratur. (Female and male in preschool).

Carlgren, I. (1995). *Professionalism som reflektion i lärares arbete, Lärarprofessionalism - om professionella lärare.* Stockholm: Lärarförbundet. (Professionalism as reflection in teachers work).

Fisher, U. & Madsen, B. L. (1984). *Titta här! En bok om barns uppmärksamhet.* Stockholm: Almqvist & Wiksell. (Look! A book about children's attention).

Garvey, C. (1990). *Play, the developing child.* (2 ed.). Cambridge, Massachusetts: Harvard University Press.

Kihlström, S. (1995). *Att vara förskollärare: Om yrkets pedagogiska innebörder.* Göteborg: Acta Universitatis Gothenburgensis. (Being a preschool teacher:on the pedagogical meanings of the profession).

Kveli, A. (1994). *Att vara lärare.* Lund: Studentlitteratur. (Being a teacher).
Lindqvist, G. (1995). *Lekens estetik, en didaktisk studie om lek och kultur i förskolan.* Forskningsrapport. Samhällsvetenskap, 95:12. Högskolan i Karlstad: SKOBA. (The aesthetics of play).
Lpfö98. (1998). *Läroplan för förskolan.* Utbildningsdepartementet. Stockholm: Fritzes. (National curriculum for preschool).
Lumholdt. (1998). *Man - mes eller macho.* Förskolan, 7. (Male - funk or macho).
Löfdahl, A. (1998). *Förhållningssätt till barns lek.* Högskolan i Karlstad: Institutionen för utbildningsvetenskap, (Attitudes to children's play).
Olofsson, B. K. (1987). *Lek för livet.* Stockholm: HLS Förlag. (Play for life).
Rasmussen, T. H. (1993). *Den vilda leken.* Lund: Studentlitteratur. (The chaotic play).
Rolf, B., Ekstedt, E. & Barnett, R. (1993). *Kvalitet och kunskapsprocess i högre utbildning.* Nora: Nya Doxa. (Quality and knowledge processes in higher education).
Strander, K. (1997). *Jag är glad att jag gick på dagis: Fyrtio ungdomar ser tillbaka på sin uppväxt.* Stockholm: HLS-förlag. (I'm happy I went to preschool. Forty youths looking back on their growth).
Sutton-Smith, B. (1997). *The ambiguity of play.* Cambridge, MA: Harvard University Press.
Sutton-Smith, B. (1999). Evolving a consilience of play definitions: Playfully. In S. Riefel (Ed.), *Play and Culture Studies, 2*, pp. 239-256). Stamford, CT: Ablex.
Wood, E. & Attfield, J. (1996). *Play, learning and the early childhood curriculum.* London: Paul Chapman Publishing Ltd.
Vygotskij, L. (1980). Leken och dess roll i barnets psykiska utveckling. In L.-C. Hydén (Ed.), *Psykologi och dialektik. En antologi i urval av Lars-Christer Hydén .* Stockholm: P A Norstedts och söners förlag. (Play and its role in children's mental development).

Part IV

The Variabilities of Adult Leisure

Finally we come to a series of chapters on adult play and have the good luck that by and large they are about the modern diversities of play. They are in a sense about the adult "playground" whereas most of the chapters in the prior sections were about the child "playroom," not about the child playground. Those are, however, legitimate foci and are clearly modern attempts to bring an adult-child play systemic culture alive within the classrooms. In some cases, they might be thought to reproduce in the classroom what were once the culture play conformities of whole societies.

CHAPTER 11. ROBYN M. HOLMES: WORKING TO PLAY: COLLEGE ATHLETES' CONCEPTIONS OF PLAY AND WORK

A constant problem in play theory is known where play stops and work begins. There is so much professional play in the modern world not only in sports, athletics, new games and X-games but also in television comedians and hosts, and more oddly with clowns, collectors, film stars, actors, artists and novelists and so on. Their performance content could be called play and often their attitudes are playful, but it is also their work. So what should we do? Many modern child psychologists have decided for themselves that play is only what children do and all the rest is something else. But this doesn't solve the problem and neglects the history and anthropology of play which shows that in many earlier societies play was thoroughly mixed into other types of activity, religious, economic and political and was not in fact distinguished from them even though having all the characteristics that you and I in our own time would like to nominate as play. (Drewal 1992; Goodman, 1992.) What this leads to introspectively is to realize that even when we are involved in what we define as play our minds are often somewhere else, so that it becomes sensible psychologically to realize that in any given so called play event there can be percentages of play and percentages of work or whatever. Play is seldom purely play all the time. Even with children, much of the play times is actually getting ready for the play, and is also often interrupted by disputes or by side exchanges of one sort or another. At this point, then, we arrive at Holmes' chapter which shows that her group of University athletes who are paid for their athleticism make subtle distinctions between the parts of their activity that are full of fun and playful (the actual games) and the parts that are work (the strenuous or routine practices). In sum while there are some who advocate "pure play" as the only kind worthy of our attention, it is very clear that that is a rarified stance which doesn't match the way things are playfully for most people most of the time (Carse, 1986).

CHAPTER 12. FELICIA McMAHON: 'COMMUNITY PLAY' AT THE MALL.

The interpenetration of play into other forms of institutional life is illustrated in an even more complex way in the present treatment of the mall as a kind of festival within American Economic Life. McMahon shows that there are multiple parallels between festivals in their traditional sense (Mardi Gras, Burning Man, July 4th) and the events that take place in malls. As in the case of the paid athletes in the prior study, likewise, the people who are in the Malls are going to vary enormously in whether they feel they are at play or not. Her article mainly illustrates the presence of 'festival- like' elements in Malls such as that they are, in the first place, a 'special' time apart or time out of time (Falassi, 1967), most obviously when there can be costumes, music, clowns, cheerleaders, restaurants, bands, film stars, and fireworks. But most importantly, the mall is a multifarious form of every changing commercial shopping stimulation that the players can watch, wander, browse, wait or wish, all states of semi indolence. Rojek (1995) calls this a state of flanerie, and the flaneur is described as a constant seeker of impressions, a condition most facilitated by the stimulus multiplicities of the festival, the state fair, the city center and now the mall. In a way, big city life as in Times Square New York has a similar ménage of alternative spectacles and entertainments which provide the visitor at least with a sense of enjoyable vicarious play. Note that in this, as in the prior study, play is sometimes most obvious and sometimes more subtle in the midst of various other orientations equally important to the event at hand. But most importantly is the way that American businesses have recognized the power of what McMahon calls "festive vocabulary" and have co-opted festival elements in order to attract customers to the mall.

CHAPTER 13. CHERYL R. H. MEAKINS, ANITA C. BUNDY, & JEFFERY GLINER: VALIDITY AND RELIABILITY OF THE EXPERIENCE OF LEISURE SCALE (TELS)

What is at first interesting about this chapter is the number of studies reviewed indicating that the experience of appropriate leisure reduces the presence of stress. Given that we favor a theory of play as a reconciliation of conflicting involuntary and voluntary emotions, these are, for us, favorable outcomes. But of greater importance, and continuing the foci of the two prior chapters, is again the great varieties of playfulness, which are a part of the leisure scale being developed by these authors. In the present case, however, we have moved into a more subjective vein asking about the feelings and behaviors of the players that define their actual play participation. It needs to be stressed at this moment that this modern choice for the study of play's subjectivity is itself an outcome of the history of play that we have been tracing through the prior sections, which is from social context to personal context centered bases for play. Here are some of the subjective items of play that the creators of this scale include. Under their headings of play there are as examples: measures of degrees of intense involvement, making ones own decisions, feeling physically and emotionally safe, demonstrating real exuberance by giggling, laughing, singing, and shouting, overcoming difficult odds, changing the play to increase the fun or the challenge, teasing and doing mischief to others, carrying on for the sheer pleasure of it, pretending to be someone else, incorporating others in your imaginings, negotiating when necessary, playing the most appropriate roles, helping others to enjoy it, telling jokes and exaggerating one's behavior.

This is a good scale and as we have said, brings us to where play needs to be in order to understand the present state of the modern culture-play subjective leisure systemic. It is an exemplification of multiplicity of play that needs to be better understood by the contributors to the kinds of narrower school play socialization outlined earlier.

CHAPTER 14. PAOLA DE SANCTIS RICCIARDONE: COLLECTING AS A FORM OF PLAY

This brings us finally and almost strangely to collection as a form of play which most of us might not have considered a very playful or modern form. And historically it may not have been, but as presented in this chapter, collecting has always been a most remarkable and highly ingenious representation of the playfulness of the collectors. What is most exciting are the illustrations of the ways in which throughout history the collectors played with the boundaries of the value of their materials as useful or useless, trivial or valuable, sane or dark, orderly or disorderly, gaming or a games, improvised or theoretical, deep or superficial, open ended or categorical, turbulent or ritual, and finally are a mockery. This is just mention a few of the labile dualities perceived and created by this author. As the author says and demonstrates in this chapter "the sense of wonder and playfulness in collecting is still alive and well."

Chapter 11

Working to Play: College Student Athletes' Conceptions of Play and Work

Robyn M. Holmes

The scholarly inquiry into sport has been explored by diverse disciplines and traversed several domains and points of the life cycle. For example, there are numerous literature bases for the developmental stages of children's, adolescents' and adults' participation in sports and sporting contests. In addition to age divisions, other aspects of sport such as consequences of sport participation on the player (e.g., Upthegrove, Roscigno & Charles, 1999), sport as a socializing agent (e.g., Coakley, 1996), motivation to participate (e.g., Buonamano, Cei & Mussino, 1995; Martindale, Devlin & Vyse, 1990), athletes' performances (e.g., Seggar, Pederson, Hawkes & McGown, 1997; Williams, Franks & Lester, 2000), emotional expression (e.g., Edwards, 1995), and the relationship between sports and culture (Dyck, 2000; Patterson, 1999; Hargreaves, 1986; Harkness, 1988; Oliver, 1986; Tomlinson, 2000) have also been explored.

Recent anthropological explorations in the relationship between games, sports and culture have focused upon cultural, social, and contextual factors (e.g., Dyck, 2000). For example, Lithman (2000) takes a Vygotskian approach to children's sport in Sweden by examining how children and parents use sport to construct and extract meaning in this particular cultural context which is embedded in a larger framework that addresses the place of sport in society. In a related work, Harris (1980) explored the relationship between cultural competency, play, and sport. She suggested that the flexibility inherent in play can be applied to cultural and individual processes and competencies. Play in sports is discussed within this framework.

The relationship between sport and socialization parallels that of sport and culture. In these instances, sport, like play, is a reflection of cultural values and beliefs and sports can function as a socializing agent (see e.g., Mitchell, 1991). For example, Coakley (1996) explored the relationship between sport and social development. He discussed examples to highlight this relationship that include moral development, community socialization, and personal and sex identity formation. This work also views sport participation as a social and interactive process dependent upon the context in which it occurs.

Play researchers have also explored the relationship between sports and other activities. In a theoretically based work, Sacks (1977) pursued a distinction between play, sport, and games by challenging existing game classification paradigms and employing Huizinga's (1955) definitional criteria for play. He suggested that sport typologies need to be expansive enough to include the fact that sports can be experienced as both play and work under certain contextual factors and when certain criteria are present. In particular, he noted that two continuums, play/work and non-athletic/athletic, would be most useful in this quest. Using these in concert, one could distinguish between activities that simply required some physical skill from those that were physically draining and activities that were extrinsically motivated, obligatory, and involuntary from those that were intrinsically motivated and voluntary. In this view, professional sports would be classified as work whereas a Saturday afternoon golf game would be considered play.

In a related work, Iso-Ahola (1980) noted that the boundaries between play and work become blurred when it comes to children's organized sports. In this study on Little League (see also Fine, 1987), pressure from parents and coaches deflates the fun kids have playing and the valuable lessons they learn from sports participation. This produces an unenjoyable and anxiety producing experience instead of a pleasurable one. Pressure to succeed in sports frequently escalates when sports participation moves from childhood to intercollegiate athletics. Parental and self-created pressure, financial pressure for receiving and maintaining scholarships, TV visibility, funding, and winning may serve either independently or collectively in influencing the student athlete's sport experiences (e.g., Anderson, 1996). At some point one could argue that playing under certain contexts is no longer fun; rather, it can become work.

This article intends to revisit the relationship between play, work, and sport. First, it expands upon the earlier work of Sacks' (1977) theoretical sports typology insofar as it is an empirical study. Second, it takes a Vygotskian approach to explore how interactions and experiences assist the players in the constructions of play and work. Emphasis is placed upon the sociocultural context in which interactions and activities occur and the meaning players create, extract, and derive from interactions as they construct their notions of play and work (Bloch & Adler, 1994; Lithman, 2000; Roopnarine, 1998).

In particular, Sacks (1977) used the example of football to illustrate how sport could be categorized as both play and work. On the play/work continuum, small time college football was placed near play whereas professional football was viewed as work. Big-time college football occupied a position somewhere between the two. The implication is that sports in small, post secondary institutions are play because they lack material interest, (e.g., player salaries or funding). However, if one considers sports as work by virtue of material interest and extrinsic motivation, then a micro-level analysis might also be fruitful. For example, in each case (small and big time college and professional sports) the players can be paid to play. In professional sports, one receives a salary. In college sports, scholarship monies can be viewed as pay for play. Thus student athletes' personal experiences such as whether they receive scholarship monies or financial aid to play would also be important to consider when one is classifying sport along a play/work continuum, and this is the issue which is explored here. Several prolegomena matters on the play/work dichotomy should be mentioned.

First, most if not all individuals involved in the study of play agree that it is an activity engaged in by both animals and humans and one that has evolutionary significance (e.g., Fagen, 1995; Powers, 2001; Sutton-Smith, 1997). However, producing a precise definition of play has been difficult (Holmes, 1999; Sutton-Smith, 1999). What seems clearer are the criteria that characterize an activity as play, and these have been duly reported in the literature.

Contemporary play researchers have pursued more accurate definitions of play by refining and expanding Huizinga's (1955) seminal criteria. Currently, play is perceived as an activity that is fun, voluntary, intrinsically motivated, framed in pretense, and concerned with process rather than outcome (Johnson, Christie & Yawkey, 1994). Sutton-Smith (1999) has suggested that adaptive variability and flexibility also be considered as critical criteria (see also Smith & Vollstedt, 1985).

Second, the play/work dichotomy has its origins in Western religious and cultural ideologies (Overman, 1983). The view of play as sinful and frivolous and hard work as praise worthy and leading to salvation stems from our Puritan ancestors and continues to influence American views and attitudes toward play (Hughes, 1999). Thus work and play presumably possess opposing characteristics or qualities. Other researchers have suggested that work is not the antithesis of play. Rather, work's position might be better occupied by leisure (e.g., Blanchard, 1986) or depression (Sutton-Smith, 1997).

Third, several researchers have challenged the mutually exclusive categorical boundaries of play and work by noting that these boundaries often overlap (e.g., Abramis, 1990; Bordin, 1979; Goodman, 1994; Kneidek, 1996; Ross, 1993; Zeigler, 1976). For example, Schwartzman (1978) highlighted the ambiguous boundaries between play and work. She suggested one could experience fun at work and find it enjoyable whereas play sometimes can be externally motivated and experienced as unenjoyable. Sutton-Smith (1997) has frequently noted that not all play is fun. At times it can be brutal, manipulative, exploitative, and cruel (see also Kelly-Byrne, 1989).

Fourth, several perspectives have contributed to the play/work literature. For example, early childhood educators have produced a great deal of literature on the place of play and work in the curriculum (e.g., Cunningham & Wiegel; 1992; King, 1982; Marshall, 1994; Perlmutter & Burell, 1995; Polito, 1994; Robson, 1993; Rosario, 1992; Wing, 1995). In addition, developmental perspectives have also generated numerous studies on the constructs play and work. For example, Holmes (1999) explored developmental trends in play and work with cross-sectional sub samples of kindergarten and college students.

The play/work dichotomy has also been pursued cross-culturally (e.g.,

Bloch, 1989; Bloch & Wichaidt, 1986; Holmes, (2001); Lancy, 1980; Nam, 1992). For example, Bloch and Adler (1994) examined the relationship between play and work and the gender differences that emerge in these activities. The study is framed in a "cultural-ecological model," and attention is directed toward the relationship between early experiences in work and play that a culture provides for its children and how these impact upon future experiences as adults (p. 150).

Fifth, the interaction of contextual factors and the play/work dichotomy have also been explored (Bloch & Pellegrini, 1989; Katz, 1986; Reifel & Yeatman, 1993). Several researchers (e.g., Bloch & Adler, 1994; Roopnarine, 1998) have emphasized the importance of attending to the sociocultural contexts in which interactions occur and how such processes direct development. For example, Western paradigms may not be appropriate for interpreting and explaining children's play and work activities in different social and ecological settings (Roopnarine, 1998). The importance of cultural context and the enculturation process as they appear along the play/work dichotomy emerge in Packer's (1994) work with kindergarten children (see also Lancy, 1996) and Wenger's (1989) time allocation study on children's play and work activities in the Kenyan community, Kaloleni.

The current study takes a particular sport experience, namely collegiate sports and examines the impact of collegiate sports participation on student athletes' conceptions of play and work. The primary purpose is to determine the definitional criteria student athletes assign to play and work. The secondary goal is to understand how the experience of playing a collegiate sport (with and without receiving scholarship monies) guides these perceptions (e.g., Sacks, 1977). It is anticipated that sports and particular aspects of sports participation will be perceived as either work and/or play because of their characteristics and qualities. At times, the categories will blur depending upon the context of the activity. The present work plans to broaden our knowledge on perceptions of play and work by exploring how one's life experiences and sociocultural contexts (i.e., participation in the subculture of collegiate sports) guide one's notions of these constructs.

METHOD

Participants

The participants were 112 student athletes from a small, private, 4-year university which has a Division I NCAA sports classification. There were 56 females and 56 males. Ethnic composition was 89 European American, 10 African American, 6 Latino and 7 student athletes from other ethnic groups. The age range was 18-23 years with a mean age of 19.50 years (SD = 1.11years). Class year statistics were: 49 freshman, 36 sophomores, 24 juniors, and 7 seniors. Fifteen participants received full financial aid (all basketball), 55 partial aid (all other sports), and 42 no aid. Eighty-four participants played only 1 sport, 28 played two, and 4 played two or more. Fifty-three participants have played for one year, 35 for two years, 20 for three years, and 4 for five years. The total sample can be divided into sub samples by athletic team. In the track and field there were 34 males (26 European American, 2 African American, 2 Latino, and 4 from other ethnic groups) and 23 females (16 European American, 4 African American and 3 Latino); in women's basketball there were 6 females (3 European American, 2 African American and 1 from other ethnic groups); in men's basketball there were 8 males (6 European American and 2 African American); in men's soccer there were 14 males (13 European American and 1 from other ethnic groups); in women's soccer there were 16 women (15 European American and 1 from other ethnic groups); and for women's field hockey there were 11 women (10 European American and 1 Latino). All participants were treated according to the American Psychological Association's (1990) guidelines for the ethical treatment of human participants.

Design and Procedure

The current study employed qualitative methods in the form of written questionnaires. This measure sought to elicit the participants' perceptions of play and work and their emotional experiences playing a collegiate sport. The questionnaire included 10 open-ended and 4 questions arranged on a Likert scale. A pilot study with 50 participants was employed to determine readability, construct validity, and completion times.

Initial consent to pursue the project was granted by the university's athletic director. Next, a research proposal was submitted to my departmental review board. After approval, I met with all the participating coaches and made arrangements to meet with each of their respective teams. Teams that participated were in their off-season (spring 2001). Men's baseball and women's lacrosse participated in fall 2001, however, these two teams were not included in this study.

I met with each team individually, briefly explained the project and my reasons for wishing to pursue it, and asked for their help in doing so. I administered all questionnaires with the exception of the field hockey team. In this case, the coach administered the questionnaires due to time constraints. All participants signed informed consents and were aware that their participation was voluntary and that they could withdraw at any time. All volunteered to complete the questionnaire and those enrolled in introductory psychology courses received course credit. All of the team questionnaires were administered in an athletics class/meeting room with the exception of track and field. This occurred in the gymnasium foyer due in part to the large number of student athletes. Participants were assured their responses would remain anonymous, received directions on how to complete the questionnaire, were told there were no right or wrong answers, and thanked for their cooperation. They completed the task in one sitting and completion times averaged about one-half hour. I remained present in the room while they completed them. Since debriefing was not possible (the questionnaires were administered prior to practice) they were given my phone number and encouraged to contact me if they had any questions. I also informed them that they could inquire about the preliminary findings at any time.

Coding

I taught a student assistant how to code the responses for the open-ended questions and served as the second coder. Categorical domains were formed from verbatim responses on the questionnaire and similar responses were subsumed into their respective domains. For example, "fun," "fun by myself," and "fun with a group of friends" were included in the larger domain "fun." For every question, multiple phrases were coded and counted separately in the final frequency counts. There were no inter-coder disagreements.

RESULTS

As expected, for this sample of respondents play was defined as "fun" and "enjoyable." These criteria were reported respectively 50.8% and 50.0% of all total responses. Gender differences arose for the criterion of "fun." Males listed "fun" 57.1% whereas this appeared 44.6% for females. Interestingly, play was defined as "participation in a game, sport, or activity" equally by males and females (16.1%) and play was viewed as "competitive." Males reported this criterion 10.7% and females did so 7.0% of the time. Other criteria worthy of mention were "action/physical movement" and "leisure activities." "Action/physical movement" emerged in 8.8% of the total responses. Again, a gender difference emerged. Females reported this criterion 12.5% whereas males did so only 5.0%. Males and females equally reported play as "leisure." This occurred 7.1% of all total responses (see Table 11.1).

Table 11.1
Question 1. Please define the word play.

	Frequencies for Males			
Responses	**Soccer**	**Basket**	**T & F**	**Total**
Enjoyable	7	5	17	29
Fun – Solo or group	6	3	23	32
Participation in an activity/game/sport	4	4	1	9
Action/Physical Movement	1	0	2	3

Continued on next page

Table 11.1 – *Continued*

Exercise	1	0	0	1
Leisure Activity	0	0	4	4
Satisfying	1	2	0	3
Voluntary	2	0	1	3

	Frequencies for Females			
Responses	**Soccer**	**Basket**	**T & F**	**Field Hock**	**Total**
---	---	---	---	---	---
Enjoyable	7	2	13	5	27
Fun – Solo or group	5	3	13	4	29
Participation in an activity/game/sport	4	0	1	4	9
Action/Physical Movement	1	1	2	3	7
Exercise	0	1	2	2	5
Leisure Activity	0	0	3	1	4
Satisfying	0	0	0	1	1
Voluntary	1	0	0	0	1

Note: Frequency totals contain the actual number of responses. Some responses contained multiple category codings. These were tallied as separate responses.

Some duplicate responses to Question 1 appeared when respondents were asked to list adjectives to describe play. However, these contrasted with the frequencies and percentages that appeared in responses for defining play. For example, 81.2% of all respondents noted the criterion of "fun" (83.9% males, 78.6% females), whereas only 29.5% reported the criterion of "enjoyable" (23.2% males, 35.7% females). Although females described play as more "enjoyable" than males did, neither sex listed this adjective with the same frequency when asked to define play.

Other prominent adjectives to emerge were "exciting," 'active," "difficult," and "competitive." The adjective "exciting" was reported equally to "enjoyable." Females listed this adjective (32.1%) slightly more than males did (26.8%). Interesting gender differences emerge for "difficult" and "competitive." For females, play was described as "difficult" 16.1% of the time whereas males did so only 1.8%. Finally, females also described play as "competitive" (8.9%) slightly more than males did (3.6%). Males and females listed "active" in similar percentages (see Table 11.2).

Table 11.2
Question 2. Please list adjectives to describe the word play.

Frequencies for Males

Responses	Soccer	Basket	T & F	Total
Fun	15	4	28	47
Enjoyable	2	3	8	13
Exciting	2	1	12	15
Difficult/tough	1	0	0	1
Laugh/humor	0	0	0	2
Leisure/relaxing	1	0	3	4
Active/physical movement	2	1	2	5
Competitive	1	0	1	2
Joyful/happy	2	0	0	0
Intense/focused	0	0	0	0
Energetic	1	0	0	1
Work	0	1	1	2
Challenging	1	1	1	3

Frequencies for Females

Responses	Soccer	Basket	T & F	Field Hock	Total
Fun	13	2	17	12	44
Enjoyable	7	1	3	9	20
Exciting	6	0	8	4	18
Difficult/tough	2	1	3	3	9
Laugh/humor	0	1	3	1	5
Leisure/relaxing	1	0	4	0	5
Active/physical movement	0	2	1	3	6
Competitive	3	0	0	2	5
Joyful/happy	1	0	2	2	1
Intense/focused	0	0	0	4	4
Energetic	0	0	1	2	3
Work	1	0	0	0	1
Challenging	0	0	0	0	0

Note: Frequency totals contain the actual number of responses. Some responses contained multiple category codings. These were tallied as separate responses.

When asked to list activities considered 'play,' respondents overwhelmingly reported "sports." Sports in general and specific sports appeared respectively 13.4% and 63.4% of the time. Gender differences emerged in the reporting of specific sports. This occurred 78.5% for males and only 48.2% for females. Two activities equally reported by males and females were games of chase (tag, etc.) and games in general. These were listed respectively 9.8% and 7.1% of the time. Other gender differences that emerged included "kicking games," "video games," "playing jokes," and "hanging out with friends." Females reported "kicking games" 16.1% of the time compared to males' 5.3%; males reported "video games" 10.7% of the time whereas only one female did so; only males listed "playing jokes and teasing" (5.4%) whereas only females listed "social activities" (5.4%) (see Table 11.3).

Table 11.3
Question 3. Please list an activity you would call play.

	Frequencies for Males			
Responses	**Soccer**	**Basket**	**T & F**	**Total**
Sports/General	3	3	2	8
Field Hockey	0	0	0	0
Tennis	0	0	0	0
Basketball	2	4	16	22
Football	1	1	2	4
Soccer	12	0	2	14
Track & Field	0	0	1	1
Bowling	1	0	1	2
Karate	2	0	0	0
Swimming	0	0	0	0
Volleyball	1	0	0	1
Intramural Sports	0	1	1	2
Games/General	1	1	1	3
Kicking Games	1	2	5	8
Chase Games	1	0	2	3
Video Games	0	1	4	5
Cards	1	1	4	6
Board Games	0	2	1	3
Playing jokes/Teasing	0	0	0	0
Sex/Hanky Panky	0	1	2	3

Continued on next page

Table 11.3 – *Continued*

	Frequencies for Females				
Responses	**Soccer**	**Basket**	**T & F**	**Field Hock**	**Total**
Sports/General	0	0	2	7	9
Field Hockey	0	0	0	4	4
Tennis	0	0	3	0	3
Basketball	1	3	0	1	4
Football	0	0	0	1	1
Soccer	8	0	3	0	11
Track & Field	0	0	0	1	1
Bowling	0	0	0	1	1
Karate	0	0	0	0	0
Swimming	0	0	0	2	2
Volleyball	0	0	0	2	2
Intramural Sports	1	0	1	0	2
Games/General	0	1	3	0	4
Kicking Games	3	1	4	1	9
Chase Games	2	0	0	3	5
Video Games	0	1	0	0	1
Cards	1	0	1	0	2
Board Games	0	1	1	0	2
Playing jokes/Teasing	0	0	0	0	0
Sex/Hanky Panky	0	0	0	0	0

Note: Frequency totals contain the actual number of responses. Some responses contained multiple category codings. These were tallied as separate responses.

In Table 11.4, as expected, this sample defined work as involuntary and an activity linked to its outcome. It was defined as "achievement oriented," "something one had to do," "an energy expending activity," "goal-oriented," and "involving extrinsic rewards." All respondents respectively listed these criteria 29.5%, 18.8%, 18.8%, 15.2%, and 10.7%. One slight gender difference emerged. Twice as many males (8) reported extrinsic motivation than females did (4). The blended nature of play and work surfaced in responses for this question. For example, one female participant wrote, "sometimes it can be fun" and a male participant responded, "it can be fun, it depends."

Table 11.4
Question 3. Please define the word "work."

Frequencies for Males

Responses	Soccer	Basket	T & F	Total
To achieve	7	2	6	15
With effort	4	0	3	7
Without effort	3	2	3	8
Something one has to do	4	1	7	12
Requires energy	4	3	6	13
Goal oriented/Improvement	0	0	2	9
External rewards – Money or purpose	0	2	6	8
Not enjoyable	0	0	2	2
Difficult	1	0	2	3

Frequencies for Females

Responses	Soccer	Basket	T & F	Field Hock	Total
To achieve	2	4	4	3	13
With effort	2	1	3	3	9
Without effort	0	3	1	0	4
Something one has to do	3	0	5	1	9
Requires energy	3	1	1	3	8
Goal oriented/Improvement	2	0	6	0	8
External rewards –					
Money or purpose	1	1	1	1	4
Not enjoyable	1	1	2	1	4
Difficult	0	0	1	0	1

Note: Frequency totals contain the actual number of responses. Some responses contained multiple category codings. These were tallied as separate responses.

A list of adjectives used to describe an activity as 'work' appear in Table 11.5. Far more variability emerged in comparison to Question 4 (Define work). Primary responses were "hard," "difficult/strenuous," "time consuming," and "boring." These occurred respectively 51.8%, 22.3%, 18.8%, and 11.6% for all respondents. Slight gender differences emerged for the adjective "hard." Males reported this 44.6% and females did so 58.9%. Similarly, females described work as "stressful" twice (8) as much as males did (4).

Table 11.5
Question 5. What adjectives would you use to describe an activity as work?

Frequencies for Males

Responses	Soccer	Basket	T & F	Total
Hard	9	2	14	25
Difficult/Strenuous	4	3	5	12
Time consuming	3	1	7	11
Boring	1	1	6	8
Tedious	0	0	6	6
Stressful	1	0	3	4
Not fun	3	1	1	5
Tiresome	2	0	1	3

Frequencies for Females

Responses	Soccer	Basket	T & F	Field Hock	Total
Hard	9	4	14	6	33
Difficult/Strenuous	6	1	4	2	13
Time consuming	1	1	3	5	10
Boring	0	0	4	1	5
Tedious	1	0	4	2	7
Stressful	5	0	2	1	8
Not fun	4	0	1	0	5
Tiresome	1	1	2	1	5

Note: Frequency totals contain the actual number of responses. Some responses contained multiple category codings. These were tallied as separate responses.

In Table 11.6, activities viewed as work included "homework," a "job," "school," and "practice." For all respondents these occurred 34.8%, 27.7%, 15.2%, and 10.7%. If one combines actual "class attendance" with "homework," school activities account for 50% of all responses. One slight gender difference emerged. Females listed "practice" (8) twice as much as males did (4).

Table 11.6
Question 6. Please give an example of an activity you would call work.

		Frequencies for Males		
Responses	Soccer	Basket	T & F	Total
Homework	7	2	9	18
Job	5	1	9	15
Class/School	4	0	5	9
Practice	1	0	3	4
Soccer	0	0	0	0
Basketball	1	0	0	1
Manual Labor	1	1	3	5
Cleaning	0	0	1	1

			Frequencies for Females		
Responses	Soccer	Basket	T & F	Field H	Total
Homework	4	1	7	9	21
Job	6	1	7	2	16
Class/School	2	1	3	1	8
Practice	4	1	1	2	8
Soccer	5	0	0	1	6
Basketball	0	2	2	1	5
Manual Labor	0	0	0	0	0
Cleaning	0	0	4	0	4

Note: Frequency totals contain the actual number of responses. Some responses contained multiple category codings. These were tallied as separate responses.

Responses comparing playing intercollegiate athletics and a paying job appear in Table 11.7. Overwhelmingly, both males (100%) and females (83.9%) agreed the two tasks were similar. Males and females primarily reported the time demands of playing sports, and the burden of responsibility in receiving scholarship monies. One male wrote, "Definitely, because we do it every day. We have to do it or we don't get paid (scholarship)." Interestingly, 39.2% of all responses contained mixed statements. Playing sports was both like and not like holding a paying job. Reasons cited were "it is fun" (15.1%) and "you don't get paid" (6.3%). 63% of the sample received full or partial aid and 37% received no aid. Some participants not receiving aid still viewed the activities as similar.

Table 11.7
Question 7. Would you compare being a student-athlete to a paying job?
Why or why not?

Frequencies for Males

Responses	Soccer	Basket	T & F	Total
Yes	14	8	34	56
Responsibility/Expectations/				
Obligations	7	0	0	7
Paid to do it	4	2	4	10
Time consuming/Effort	7	2	14	23
Harder than work	1	0	2	3
Do things you don't want to but				
have to	1	2	4	7
Rewards	0	0	2	2
No	2	5	16	23
Fun/enjoyable	1	2	5	8
Rewards are personal				
and not measured in dollars	0	0	2	2
Playing sports is voluntary	1	0	1	2
Benefits school	0	0	1	1
Don't get paid/Pay takes the				
fun out of it	0	0	0	0

Frequencies for Females

Responses	Soccer	Basket	T & F	Field Hockey	Total
Yes	14	7	18	6	47
Responsibility/Ex-					
pectations/Obligations	1	3	10	3	17
Paid to do it	5	1	1	0	7
Time consuming/Effort	1	3	7	1	12
Harder than work	1	2	1	0	3
Do things you don't					
want to, but have to	0	0	0	2	2

Continued on next page

Table 11.7 – *Continued*

Rewards	2	0	0	1	3
No	5	1	10	5	21
Fun/enjoyable	2	1	4	2	9
Rewards are personal and not measured in dollars	1	0	0	0	1
Playing sports is voluntary	0	0	2	0	2
Benefits school	0	0	0	0	0
Don't get paid/ Pay takes the fun out of it	0	0	4	0	4

Note: Frequency totals contain the actual number of responses. Some responses contained multiple category codings. These were tallied as separate responses.

Responses to classify playing sports along a play/work continuum appear in Table 11.8. As expected, the majority of participants (81.3%) categorized sports participation as a combination of play and work dependent upon the task and context. Responses for this query were diverse but tended to focus upon game play as fun and practice or "work to play" as work (34.8%).

Table 11.8
Question 8. Would you consider your participation in collegiate sports play, work, neither work or play or some combination of the two? Please explain.

Frequencies for Males

Responses	Soccer	Basket	T & F	Total
Play	1	0	1	2
No pay	0	0	1	1
Fun	1	0	0	1
Work	1	2	4	7
Time	0	1	1	2
Takes away the fun	1	0	0	1
Neither play or work	0	0	1	1
Not required/hard/sometimes fun				
Play and Work	11	5	29	45
Some work/play	0	1	0	1

Continued on next page

Table 11.8 – *Continued*

Fun in competing/Practice-hard	3	0	15	18
Work-time consuming; play is fun	0	0	6	6
Love the game/				
Work is commitment	2	2	0	4

Frequencies for Females

Responses	Soccer	Basket	T & F	Field H	Total
Play	1	1	1	1	4
No pay	0	0	0	0	0
Fun	1	1	1	1	4
Work	0	1	1	0	2
Time	0	0	0	0	0
Takes away the fun	0	1	0	0	1
Neither play or work	0	0	0	0	0
Not required/					
hard/sometimes fun					
Play and Work	13	3	19	11	46
Some work/play	3	0	2	0	5
Fun in competing/					
Practice-hard	7	3	6	5	21
Work-time consuming;					
play is fun	0	0	0	2	2
Love the game/					
Work is commitment	0	0	0	0	0

Note: Frequency totals contain the actual number of responses. Some responses contained multiple category codings. These were tallied as separate responses.

Participants were also asked to rate on a 5-point Likert scale (1-always play to 5-always work) aspects of their athletic experience ranging from specific elements of practice to community service. Student athletes from a variety of teams were polled for the primary responsibilities, obligations, and elements of their sports experience. The twelve items that appear in Table 11.9 were generated from that response pool. A reliability analysis was performed on the 12 items and alpha = .68. When items not related to actual sports play were eliminated (#10, 11, 12), reliability was .73. Several of the items are discussed below and unless otherwise noted refer to ratings reported by all respondents.

Table 11.9
Likert Rating Frequencies for Aspects of Sport Participation

Activity	Always Play		Sometimes Play		Neither Work or Play		Sometimes Work		Always work	
Running drills	0	**1**	7	**7**	3	**9**	23	**18**	23	**21**
Stretching	2	**4**	9	**10**	31	**31**	5	**7**	9	**3**
Weight lifting	2	**1**	13	**3**	6	**4**	13	**22**	22	**25**
Fundamental drills	1	**2**	11	**11**	16	**10**	19	**20**	9	**13**
Scrimmaging	16	**18**	15	**15**	11	**3**	5	**9**	6	**9**
Game play	21	**23**	11	**5**	9	**3**	2	**6**	13	**18**
Chalk talk/films	3	**3**	9	**9**	19	**26**	15	**15**	5	**3**
Warm-up Drills	2	**4**	14	**11**	17	**12**	14	**24**	9	**5**
Calisthenics	1	**1**	8	**7**	17	**18**	17	**18**	12	**11**
Sports psychologist	3	**10**	2	**7**	31	**31**	6	**5**	11	**1**
Community service	8	**12**	8	**14**	16	**14**	9	**12**	15	**4**
Fund raising	10	**12**	9	**16**	14	**11**	11	**14**	12	**3**

Note: Male frequencies appear in the left column; female frequencies appear in bold in the right column.

Aspects of practice pertaining to conditioning and involving exertion were perceived as work. For example, running drills received almost equal ratings for sometimes work and always work. This occurred respectively 36.6% and 39.3% of all responses. Similar percentages emerged for weightlifting. However there was a gender difference. Males rated this activity as sometimes play (23.2%) far more than females did (5.3%).

By contrast, non-exertive activities such as stretching were rated as neither work nor play. This was reported 55.4% by all respondents. Similar ratings appear for chalk/talk/films. Respondents rated this activity as neither work nor play 40.2% more so than other responses. Similar ratings emerged for visits with the sports psychologist. Respondents rated this activity as neither work nor play 55.4% of the time. However gender differences emerged. Males rated this activity as always work (19.6%) whereas females did so only 1.8%.

Warm-up drills also yielded interesting results. Responses for this element of practice produced almost equal percentages in all categories except for the end points. They were perceived as either sometimes work, sometimes play, or neither work or play. Ratings for calisthenics mirrored those of warm-up drills.

Scrimmaging and actual game play were perceived as play activities. All respondents rated scrimmaging as always play 30.4% and sometimes play 26.8% and rarely rated as always work 13.4%. Game play received a greater percentage of responses for always play 39.3% yet also was rated as always work 27.7%.

Obligations that were part of the experience but not related to sports participation were community service and fund raising activities. Each produced similar results. Four of the five ratings were reported in almost equal percentages (approximately 22.3%).

DISCUSSION

For this sample, it seems reasonable to suggest that the experience of participating in intercollegiate sports guides one's perceptions of play and work and that certain aspects of sports participation are classified along the play/work continuum by virtue of their characteristics and qualities. For example, this sample defined play as an activity that is fun and enjoyable. These are two qualities that consistently emerge in theoretical (e.g., Johnson, Christie & Yawkey, 1994) and empirical studies (e.g., Holmes, 1999) on play. However, play for this sample also includes the act of participating in a game, sport, or activity and is competitive and physical. When asked to list play activities, the majority of the responses were either sports in general or specific sports. The inclusion of competition, physicality, and sports as play preferences is most likely linked to their participation in sports and athletic experiences (e.g., Anderson, 1996).

For this sample, work is defined as something one has to do and a task that is goal-oriented, extrinsically motivated, and focuses upon outcome. These criteria also are reported elsewhere in the literature (e.g., Huizinga, 1955; Johnson, Christie & Yawkey, 1994). In addition, work is reported to be time-consuming and hard. One might suspect that time-consuming appears by virtue of the time demands placed upon student-athletes because they are required to perform well in the classroom and on the field, both of which require time and commitment. Thus it is logical that this sample reports schoolwork as its primary response for a work activity because they are also members in the sociocultural world of college. Similar findings for other participants in various school cultures have been reported in the literature (e.g., Holmes, 1999).

As expected, this sample likens playing a college sport to a paid position, hence work. Primary reasons include time demands and receiving athletic scholarships. What emerges from the findings is a sense that these college athletes realize the responsibility of being paid to play and consider their participation as work. In this case, personal funding guides one's perceptions of playing a college sport along the play/work continuum. For example, one female participant wrote, "A job and being a student-athlete both incorporate work and play (sometimes) and if you are on a scholarship it is like a paying job because it is paying for your education." In Sacks' (1977) model, institutional funding and salaries figure prominently as material interest thus placing small-time college athletics closer to the polar end of play. However if one considers the player's perspective and personal material interest, then it seems reasonable that student-athletes receiving aid regardless of the size or ranking of the institution would consider playing a collegiate sport work. Thus exploring a player's perspective will help broaden our typology of sports.

Also, this sample clearly illustrates the blurred boundaries of play and work. First, although most participants compared sports participation to work, approximately 40% of the responses contained mixed statements. It could also be play because "it is fun" and "you don't get paid." One male participant noted, "you work to play" and one female participant viewed sports participation as "work with some play. We do so much of it, the fun is often taken away and it turns into work."

As expected, student athletes' receiving scholarship monies all responded that playing their sport was comparable to a job. Second, when asked to classify playing collegiate sports along the play/work continuum the majority of the participants viewed their experience as a combination of play and work dependent upon task and context. For example, game play was considered play yet practice was viewed as work.

Finally, certain aspects of one's sport are classified along the play/work continuum by virtue of their characteristics and qualities. For example, aspects of training that involve physical exertion such as running drills were viewed as work whereas benign activities such as stretching, viewing films, and visiting the sports psychologist were viewed as neither work or play. Play included game play and scrimmaging, which this sample often viewed as fun.

There were some limitations to the present study. First, this sample was drawn from a small, private university and was relatively homogeneous sample with respect to ethnicity. Second, only seven sports were represented with some sports having a larger sub sample than others. For example, women's basketball only had 6 participants. Challenges to external validity are acknowledged and I would be reluctant to extend these findings to a broader sample.

In conclusion, it seems reasonable to suggest that the experience of playing intercollegiate sports guides one's notions of play and work and how one classifies activities along the play/work continuum. What emerges from these findings is the importance of considering the collegiate sport experience from the player's perspective. It seems clear that student athletes who receive scholarship monies realize their responsibilities. This is reflected in their responses that playing a sport is akin to being paid for a job. For example, one male participant wrote, "being on scholarship, I am basically being paid to play soccer. So soccer is my job."

This study could have both theoretical and applied value. First, sports typologies could be expanded to include the player's perspective (see e.g., Sacks, 1977). Second, athletics departments could benefit from learning how the players experience and process their sports participation. Because student-athletes are susceptible to a host of problems (e.g., Anderson, 1996) such explorations could provide information on student-athlete welfare and well-being. Future explorations might examine gender and ethnic differences, amount of received scholarship dollars and consider a cross-sectional study involving players from diverse institutions and NCAA divisions.

Note

1. The author sincerely thanks the student athletes, coaches, athletic director, and athletics department for their enthusiastic participation in this project, her student, Janet Gallo for coding the questionnaires, and her qualitative methods students. Correspondence may be addressed to: Dr. Robyn M. Holmes, Dept. of Psychology, Monmouth University, West Long Branch, NJ 07764-1898 USA; e-mail: rholmes@monmouth.edu.

REFERENCES

Abramis, D. (1990). Play in work: Childish hedonism or adult enthusiasm? *American Behavioral Scientist, 33,* 353-373.

American Psychological Association. (1990). *Ethical principles in the conduct of research with human participants.* Washington, DC: Author.

Anderson, M. (1996). Working with college student-athletes. In J. Van Raalte & B. Brewer (Eds.), *Exploring sport and exercise psychology* (pp. 317-334). Washington, DC: American Psychological Association.

Blanchard, K. (1986). Play as adaptation: The work/play dichotomy revisited. In B. Mergen (Ed.), *Cultural dimensions of play, games, and sport* (pp. 79-88). Champaign, IL: Human Kinetics Publishers.

Bloch, M. (1989). Young boys' and girls' play at home and in the community: A cultural-ecological framework. In M. Bloch & A. Pellegrini (Eds.), *The ecological context of children's play* (pp. 58-74). Norwood, NJ: Ablex Publishing Co.

Bloch, M. & Adler, L. (1994). African children's play and the emergence of the sexual division of labor. In J. Roopnarine, J. Johnson, & F. Hooper (Eds.), *Children's play in diverse cultures* (pp. 148-178). Albany, NY: State University of New York Press.

Bloch, M. & Pellegrini, A. (Eds.) (1989). *The ecological context of children's play.* Norwood, NJ: Ablex.

Bloch, M. & Wichaidt, W. (1986). Play and school work in the kindergarten curriculum: Attitudes of parents and teachers in Thailand. *Early Child Development & Care, 24,* 197-215.

Bordin, E. (1979). Fusing work and play: A challenge to theory and research. *Academic Psychology Bulletin, 1,* 5-9.

Buonamano, R., Cei, A. & Mussino, A. (1995). Participation motivation in Italian youth sport. *Sport Psychologist, 9,* 265-281.

Coakley, J. (1996). Socialization through sports. In O. Bar-Or (Ed.), *The child and adolescent athlete* (pp. 353-363). Oxford: Blackwell Scientific Publications, Inc.

Cunningham, B. & Wiegel, J. (1992). Preschool work and play activities: Children and teacher perspectives. *Play & Culture, 5,* 92-99.

Dyck, N. (Ed.). (2000). *Games, sport and culture.* New York: Berg.

Edwards, S. (1995). Edwards Inventory of Emotions: Assessing emotions in athletes and non-athletes. *Perceptual & Motor Skills, 80,* 444-446.

Fagen, R. (1995). Animal play, games of angels, biology and brain. In A. Pellegrini (Ed.), *The future of play theory* (pp. 23-44). Albany, NY: State University of New York Press.

Fine, G. (1989). *With the boys.* Chicago: University of Chicago Press.

Goodman, J. (1994). "Work" versus "play" and early childhood care. *Child & Youth Care Forum, 23,* 177-196.

Hargreaves, J. (1986). *Sport, power, and culture.* Cambridge, UK: Polity Press in association with Basil Blackwell.

Harkness, D. (1988). *Sports in American culture.* Tampa, FL: American Studies Press.

Harris, J. (1980). Play: A definition and implied interrelationships with culture and sport. *Journal of Sport Psychology, 2,* 46-61.

Holmes, R. (2001). Parental notions about their children's playfulness and children's notions of play in the United States and Hong Kong. In S. Reifel (Ed.), *Play & Culture Studies, Vol. 3: Theory in Context and Out* (pp. 291-314). Westport, CT: Greenwood Press.

Holmes, R. (1999). Kindergarten and College Students' Views of play and work at home and school. In S. Reifel (Ed.), *Play and Culture Studies, Vol. 2: Play Contexts Revisited* (pp. 59-72). Stamford, CT: Ablex Publishing Corp.

Hughes, F. (1995). *Children, play, and development.* Boston: Allyn & Bacon.

Huizinga, J. (1955). *Homo ludens: A study of the play element in culture.* Boston: Beacon Press.

Iso-Ahola, S. (1980). Who's turning children's Little League play into work. *Parks and Recreation, 15,* 51-54,77.

Johnson, J., Christie, J. & Yawkey, T. (1994). *Play and early child development.* Glenview, Ill: Scott, Foreman.

Katz, C. (1986). Children and the environment: Work, play and learning in rural Sudan. *Children's Environments Quarterly, 3,* 43-51.

Kelly-Bryne, D. (1989). *A child's play life: An ethnographic study.* New York: Teachers College Press.

King, N. (1982). Work and play in the classroom. *Social Education, 46,* 110-113.

Kneidek, T. (1996). The early years: When work is play. *Northwest Education, 2,* 2-7.

Lancy, D. (1996). *Playing on the motherground: Cultural routines for children's development.* New York: Guilford.

Lancy, D. (1980). Work and play: The Kpelle case. In H. Schwartzman (Ed.), *Play and Culture* (pp. 324-328). West Point, NY: Leisure Press.

Lithman, Y. (2000). Reflections on the social and cultural dimensions of children's elite sport in Sweden. In N. Dyck (Ed.), *Games, sport and culture* (pp. 163-181). NewYork: Berg.

Marshall, H. (1994). Children's understanding of academic tasks: Work, play, or learning. *Journal of Research in Childhood Education, 9,* 35-46.

Martindale, E., Devlin, S. & Vyse, S. (1990). Participation in college sports: Motivational differences. *Perceptual & Motor Skills, 71,* 1139-1150.

Mergen, B. (1986). (Ed). *Cultural dimensions of play, games, and sport.* Champaign, IL: Human Kinetics Publishing.

Mitchell, T. (1991). *Blood sport: A social history of Spanish bullfighting.* Philadelphia: University of Pennsylvania Press.

Nam, H. (1992). Conceptions of play and work held by mothers of preschool children in Korea. *Korean Journal of Child Studies, 13,* 229-240.

Oliver, L. (1986). Implications for the soccer phenomenon in American. In B. Mergen (Ed.), *Cultural dimensions of play, games, and sport* (pp. 191-208). Champaign, IL: Human Kinetics Publishers.

Overman, S. (1983). Work and play in America: Three centuries of commentary. *Physical Educator, 40,* 184-190.

Packer, M. (1994). Cultural work on the kindergarten playground: Articulating the ground of play. *Human Development, 37,* 259-276.

Patterson, B. (1999). *Sport, society & culture in New Zealand.* Wellington: Victoria University of Wellington.

Perlmutter, J. & Burell, L. (1995). Learning through "play" as well as "work" in the primary grades. *Young Children, 50,* 14-21.

Piaget, J. (1959). *The thought and language of the child.* (M. Gabain, trans.). London: Routledge & Kegan Paul.

Piaget, J (1962). *Play, dreams, and imitation in childhood.* New York: Norton.

Polito, T. (1994). How play and work are organized in a kindergarten classroom. *Journal of Research in Early Childhood Education, 9,* 47-57.

Powers, T. (2001). *Play and exploration in children and animals.* Mahwah, NJ: Lawrence Erlbaum Associates.

Reifel, S. & Yeatman, J. (1993). From category to context: Reconsidering classroom play. *Early Childhood Research Quarterly, 8,* 347-367.

Robson, S. (1993). "Best of all I like choosing time:" Talking with children about play and work. *Early Child Development and Care, 92,* 37-51.

Roopnarine, J. (1998). The cultural contexts of children's play. In O. Saracho & B. Spodek (Eds.), *Multiple perspectives on play in early childhood education* (pp. 194-219). Albany, NY: State University of New York Press.

Rosario, L. (1992). All work and no play: cram schools keep alive education nightmare. *Far Eastern Economic Review, 155,* 21-23.

Ross, A. (1993). Playing at work? Means and ends in developing economic and industrial awareness in the early years of education. *Early Child Development and Care, 94,* 67-83.

Sacks, H. (1977). Sport: Play or work? In P. Stevens (Ed.), *Studies in the anthropology of play: Papers in memory of B. Allan Tindall* (pp. 186-195). West Point, NY: Leisure Press.

Schwartzman, H. (1978). Introductory notes to Chapter IV. In M. Salter (Ed.), *Play: Anthropological perspectives* (pp. 185-187). New York: Leisure Press.

Seggar, J., Pederson, J., Darhl, M., Hawkes, N. & McGown, C. (1997). A measure of stress for athletic performance. *Perceptual & Motor Skills, 84,* 227-236.

Smith, P. & Vollstedt, R. (1985). On defining play: An empirical study of therelationship between play and various play criteria. *Child Development, 56,* 1042-1050.

Sutton-Smith, B. (1999). Evolving a consilience of play definitions: Playfully. In S. Reifel (Ed.), *Play and Culture Studies, Vol. 2: Play Contexts Revisited* (pp. 239-256). Stamford, CT: Ablex.

Sutton-Smith, B. (1997). *The ambiguity of play.* Cambridge, MA: Harvard University Press.

Tomlinson, A. (2000). *The sports studies reader: a reader on sport, culture and society.* London: Routledge.

Upthegrove, T., Roscigno, V. & Charles, C. (1999). Big money collegiate sports: Racial concentration, contradictory pressures, and academic performance. *Social Science Quarterly, 80,* 718-737.

Vygotsky, L. (1978). *Mind in society.* Cambridge, MA: Harvard University Press.

Wenger, M. (1989). Work, play and social relationships among children in a Giriama community. In D. Belle (Ed.), *Children's social networks and social supports* (pp. 91-115). New York: John Wiley & Sons.

Whiting, B. & Edwards, C. (1988). *Children of different worlds.* Cambridge, MA: Harvard University Press.

Williams, D., Frank, M. & Lester, D. (2000). Predicting anxiety in competitive sports. *Perceptual & Motor Skills, 90,* 847-850.

Wing, L. (1995). Play is not the work of the child: Young children's perceptions of work and play. *Early Childhood Research Quarterly, 10,* 223-247.

Zeigler, E. (1976). *A meta-ethical analysis of "work" and "play" as related to North American sport.* Paper presented at the 6th Annual Meeting of the Philosophic Society for the Study of Sport. Hartford, Connecticut, October 21-23.

Chapter 12

Community Play at the Mall

Felicia Mcmahon

Viewed as a form of community play, festival is as ephemeral as a daydream. Each festival and dream is unique. At the same time, both festival and dream can be replayed endlessly. As folklorist Roger Abrahams so aptly says about festival, we build it up merely to destroy it. Each year, throughout the world, the same process is repeated (1987). Firecrackers, bonfires, costumes, games, and festival foods all carry a similar message: life is temporary, yet eternal. We celebrate it by building it up, only to consume it.

This process is apparent in American malls where businesses have co-opted festival vocabulary in order to attract consumers to "join in the celebration" and to "get in the spirit" by purchasing goods. So too Zepp (1986) argues the modern mall is a center for both physical and spiritual renewal, where Americans go to shop as well as to be entertained, relax, exercise or to attend organized religious services. (Like any public display event, it is not possible to present a complete picture of all the potential meanings that may exist in a particular festival. In this paper, however, I identify the major elements of festival in several celebrations around the world, including the mall experience, which has become a "festival experience" for many Americans. It is possible to discern the coded patterns of festival vocabulary and to demonstrate that the festival, as a form of community play, continues to be an unexamined aspect of anthropological life when, in fact, it is integral to the proper differentiation of rhetoric's about what is traditional and culturally appropriate. In addition, within American culture, the mall experience has become a festival experience for many.

CONSUMING THE FESTIVAL

In his cynical and piercing critique of consumerism, Eduardo Galeano writes, "This fiesta, this great global binge, makes our heads spin and clouds our vision; it seems to have no limits in time or space. But consumer culture is like a drum: it resonates so loudly because it's empty" (2000, p. 248). Galeano employs "festival" as a root metaphor to describe the enticements of consumerism yet he ignores the ways in which festival vocabulary is understood and employed by American businesses to entice shoppers. This may be due because, as Roger Abrahams notes,

> The marketplace itself has been overlooked as a source of significant cultural activity because commercial enterprises have been set up in opposition to cultural creativity. Perhaps this is because the marketplace as well as fairs and festivals draw upon ephemeral cultural products-consumables and decorative devices intended to be used up, blown up, thrown away-as a basic means of attracting consumers and revelers. (1990, n.p.)

One needs only to enter through the doors of any modern mall to recognize these festival devices: Immediately, we hear noise and music, and the crowds abound. There are no clocks, indicating that we have entered a "time out of time." Balloons, flags and colorful banners hang from everywhere while bright lights and tantalizing smells surround us. Salespeople approach us with free food samples or complimentary squirts of the latest perfumes. We become intoxicated by the bright sights, smell, sounds, and tastes of this "festival." Staged performances mesmerize shoppers in the walkways where we can buy chances on a shiny new car displayed in the mall or hear an announcement of a blue light sale. We dash off to get yet another bargain. Shoppers experience Victor Turner's communitas as the group's energy flows. We are encouraged to buy what is new and better, items that we may not need, objects produced only to be consumed, used up, discarded, thrown out only to seek out once again the festival experience at the mall.

THROUGH THE ANALYTICAL LENS

Although no one theory of festival explains this kind of group play or any other play form entirely, multiple theories can provide frameworks for looking at festivals from differing perspectives. For Americans, the mall experience has become a form of "community play:" leisure time is shared with friends and family at the mall where they shop, eat, play video games or take in a movie. In the anthropological literature on festival, Turner (1969) has been one of the most influential theorists. His communitas encompasses an anti-structure within a society's accepted structure. Applying this kind of thinking, we could interpret time spent at the mall with family and friends as an expression of anti-structure to the intimacy of the home. This communitas at the mall, a shared "group flowing," is comparable to sociologist Csikszentmihalyi's description of "flow" as a peak experience on the individual level (1990).

This is not to say that in festive play there is no structure; indeed, it is just an alternative structure, acceptance of which creates a bond among the participants. A part of Turner's theory incorporates the earlier works of Leach (1961) and Gluckman (1963) who interpreted festival as a "release" or "safety valve" by which groups avoid actual conflicts, which might threaten the society itself. Although for Turner there are elements of release from the accepted order of things, he sees the symbolic action of festival as dramatically structured. This is apparent in the so-called "rag-tag" costumes of clowns whose costumes are in fact designed in a specific way to give the appearance of being rag-tag. Often the "Stranger" or "It" figure in the form of an unidentified person or frightening creature is used to mobilize the crowds and represents the exclusion that is at the center of every festival.

This so-called reaffirmation of the usual structure through the anti-structure of festival is not accepted by more recent festival theorists. It is assumed that inversion is the essential element in festivals, especially in carnivals. However, gross exaggeration can just as easily replace inversion. Following Manning, "Play inverts the social order and leans toward license, whereas ritual confirms the social order and is contrastive" (7) but Manning's statement does not take into account the incorporation of play into religious rites as noted by Drewal (1992) in her work on Yoruba culture or in festivals like the Days of the Dead.

This is not to say that reversal behavior described by the literary theorist, M.M. Bakhtin (1968), as well as a host of anthropologists and folklorists such as Babcock (1978), Bauman and Abrahams (1978) and other notable scholars, is not an important element of festive play. In his study of the bawdy novel by the sixteenth century writer, Francois Rabelais (1965/1532), Bakhtin found a similarity between the novel and the carnival. Bakhtin suggests that, in the novel, the author's voice through the written text has the potential to creatively change accepted norms just as the carnivalesque choir questions authority. However, Bakhtin's view of the power of carnival to cause change is somewhat romantic: more often than not, the festival does not boil over into total revolution against society's norms because the norms themselves are a part of every festival.

Into this mix of useful yet in themselves limited theories, Schechner (1988) and Geertz (1972) throw in "deep play" as an element on both the individual and community levels. For Schechner, deep play means risky behavior. In community play, there exists danger: what would happen if the everyday order were NOT restored? Geertz, on the other hand, believed that root metaphors are expressed in play and that these metaphors can be discerned when the anthropologist looks intently "over the shoulder" of the players. She or he "reads" between the lines of the symbolic play script. Although scholars later pointed out the limitations of Geertz's theory because there is no single "meaning" for participants in any play event, Geertz never asked the participants how THEY viewed their own play.

In addition, what do participants have to say about their own play? During the Christmas holiday season, when American shoppers must encounter huge crowds and long lines to make sometimes superfluous gift purchases, it is common to see reports on American television airing comments made by shoppers about their reasons for participating in this hectic shopping tradition. "I enjoy the music and the crowds," said one older European-American male shopper. "I just love shopping," claimed a younger African-American female during the same interview. From these and many other similar statements, we can assume that the mall experience, at least at Christmas time, provides a form of community play for many shoppers. Fernandez (1986, 1991) further explores the interpretation of root metaphors in community play. For Fernandez, festivals provide an opportunity not only for the group to flow but also to create a distinctive identity, which separates outsiders from group members. We can see this as exclusion within an inclusive event in festivals such as the New Orleans Mardi Gras where Krewes or float clubs are clearly marked and limited to members only. At the same time, historian Kinser writes, "Playful as it is, richly inventive as it has been, Carnival at New Orleans and Mobile conserves the black-white barriers which can be observed nearly everywhere in American society" (1990, p. *xx*). Likewise, within the mall world, those who have money to spend can play; those who do not are excluded.

From a feminist perspective, theorists also explore the ways in which women are excluded or only marginally represented in festivals as well as the ways in which women insert themselves in festive play (McMahon, 1993b, 2000; Kinser, 1990). Often, women play "Backstage Marys" who may organize the festival but are not highly visible unless they are "represented" by men dressed as women. When women are placed in highly visible places, they take token roles where they sit on cars or floats, posing as the Beauty Queens for male-dominated groups. If women do portray women, they often take on the stereotypes that men have of women and act these out as a form of playful protest. We can view these playful protests as a kind of women's festival vocabulary. It is certainly obvious to this researcher that women are both at the center of the mall festival experience and simultaneously occupy roles in advertising displays comparable to the Beauty Queen in the parade car.

Descriptive Ethnography

Compare the following two festival descriptions, the first of Burning Man festival in Nevada; the second of Carnival in Germany:

1. "Slowly, the Man's arms rose from its sides and locked in a victory position. The crowd went rabid. A man in a Mylar suit walked to the hay platform, set himself on fire and ran up the steps. He gyrated like a self-immolated puppet for a few seconds and lighted a fuse on one of the Man's legs before he ran down the other side, and rolled around on the ground to put himself out.

First, fire blasted from the Man's feet and hands, then slowly crept up his legs, torso and finally his arms, setting off thousands of sparklers and firecrackers along the way. Two minutes after the fuse was lighted, his head exploded. Colored balls of fire rocketed upward. The Man wobbled drunkenly for a few seconds, then crashed to the ground. Catharsis.

Some people ran up and tossed handfuls of money into the flames. Others set fire to every hay bale and wooden structure in the vicinity. Stilt-walkers in evil clown costumes ran wobbly through the chaos. Two men in bunny suits watched impassively as the rocket thing spewed blue flame. A camera crew from MSNBC was chased from the area by a mob chanting "Burn the Media!"

The festival was now frenzy. I walked into the desert to calm down and get a panorama view. Hualapai Playa looked like a village sacked by

Mongol hordes. Dark silhouettes danced around dozens of fires in the distance. On the Burning Man main stage, the San Francisco galactic funk band Beyond Race rocked a manic crowd. The covered-wagon thing rolled up on a huge, wooden duck and promptly flambéed the fucker. I dropped by a tent full of strobe lights, mirrors and foam bats, then made one last pass at Bianca's Smut Shack, where a five-way sex show was well under way. Big deal. By then, my senses were cauterized. I was beyond shock. I was burned out. It was over.
(http://www.burningman.com/archive)[1]

    2. "My first encounter with the carnival tradition was in Elzach, Germany. As we walked along an unlit street on a frosty February night, out of the darkness individual shapes began to appear and expectant festival participants huddled along the narrow streets. We heard the faint sound of flutes in the distance. Everyone strained to see. We suddenly saw fiery torches held by demon-like bears hopping out of the darkness. Hundreds of red and green costumed figures appeared, their faces invisible behind wooden demon masks, their heads covered with tricorn hats bedecked with thousands of snail shells and trimmed with giant red balls. I pulled away from the torches waved in my face by the garish, red figures. My ears reverberated with the hollow thud of inflated pig bladders bound to slender switches and wielded weapon-like by the prancing figures at curbside. The effect was electrifying, and I reeled disoriented, hurled backward to some primeval consciousness.

    The masked figures called Narren (fools) bellowed to the crowd repeatedly, "Narri!" and the crowd responded loudly, "Narro" A band of clown like figures continued the music as the demons built a bonfire and burned a monstrous demonic effigy. The lilting, hypnotic music pierced the dark night air as the flame-silhouetted revelers pranced about the flames. With a studied adroitness, they leapt over the fire, soaring above it to the cheers of the clustered onlookers. This ritual continued until the flames died. Then, sweeping the ashes aside, several satyr-like creatures turned their attention to the enthused crowd and chased squealing young girls into the late-winter night. Others strode into the restaurants encouraging us to spend the evening in revelry.

    Precisely at midnight, all merry-making abruptly halted, and the crowd made its way homeward through the frigid, dark and now quiet streets. Completely stunned and emotionally exhausted I made my way back to my hotel room. My memory pleaded for an answer to the question that tumbled about in my head: What was the meaning of all of this?" (McMahon, 2000, pp. 378-9.)

Bonfires and fireworks play central roles in both festivals. These two descriptions of festive community play-one occurring in the fall and the other in winter-are similar in form and experience yet separated by 9,000 miles between the U.S. West Coast and southern Germany. Both employ festival language, most notably, the bonfire and firecrackers: "Perhaps the firecracker carries the message of festival most fully," Abrahams tells us, "As a noisemaker that demands attention as it consumes itself. And in the fireworks show, it becomes the most dramatic and temporary of all the festival arts, made for the moment of display only, destined to self-destruct, come apart, and disappear" (1987, p. 180). The two festivals are not anomalous: in fact, similar kinds of celebratory play continue to take place on a regular basis throughout the world, in every culture, although on different occasions. Such conduct in industrialized countries "flaunts" the hallowed Western work ethic, according to anthropologist Frank Manning: "The florescence of celebration in contemporary societies is truly striking," says Manning (1985, p. 4). "The celebrant takes 'time out' from practical affairs and ordinary routine and does so openly, consciously and with the general aim of aesthetic, sensual and social gratification." Manning's comment could easily be applied to the mall festival experience for shoppers.

When the organizers of any festival or community celebration are asked to give explanations for their respective festivals, the answers vary: organizers of the Burning Man festival (described above) say that the Burning Man is a celebration of the ephemeral nature of art. German organizers of the carnival tradition claim that it is an ancient custom to welcome spring and "to burn out evil winter spirits" (McMahon, 2000). These claims, however, are called "rhetorics of play" or in this case, "festival rhetoric" (Sutton-Smith, 1995) for when the actual participants in these festivals are asked the reasons for their participation, they invariably reply, "It's just fun!" This leads one to ask, what do we mean when we say something is "fun?" Obviously, all "fun" is contextualized: what is fun for one person may not be fun for someone else. In addition, the word "fun" is often associated culturally with the word "play" which some interpret as "not serious." Yet we know that some play is very serious: just consider American spectator sports such as football or the traditional Mexican fiesta, known as *Dias de Muertos*, the most important celebration in Mexico's yearly religious cycle.

In some American cities with a Chicano community, such as Los Angeles, the Day of the Dead has evolved into a fiesta involving a parade of many thousands of people wearing costumes that resemble playful skeletons. There are special foods, music and dances in the street. In Mexico it is believed that the ancestors' souls return each year for a brief time to enjoy life once again: "In the urban setting of Mexico City and other large towns the celebration is seen at its most exuberant, with figures of skulls and skeletons everywhere. These mimic the living and deport themselves in a mocking modern dance of death" (Carmichael & Sayer, 1991, p. 9). Such a festival as *Dias de Muertos* is not limited to the Chicano population: "So infectious is the spirit of the event that non-Latin areas of Los Angeles have been caught up in the exuberance. The day ends with the celebration of a mass" (Carmichael & Sayer, p. 71). Here, in the festival of the Day of the Dead, there is no separation between seriousness and fun; religion and play.

Thus, it makes it easy to criticize neat, polar opposites as "ritual vs. festival" or Durkheim's classic view of "sacred vs. secular" (1915). In every form of community play, there are many voices. The most obvious are the voice of the past "historical account," the voice of the present "participants," and the voice of the researcher of the written text. But since a phenomenon like festival is so complex, there are also varying levels of participation at any one festival and thus, "the voice" of the present consists of many differing perspectives. Compare this Burning Man participant's account of the same festival recorded above:

Today the city is restless; it's the day of the burn. People were still streaming into the gate late last night, looking for a place to camp. Apparently there are those who still feel that Burning Man is all about burning down a big wooden statue, and so they arrive at the last minute looking for a spectacle. But the real beauty of this city is more subtle. A walk through town will show you more about this community than merely attending the big bonfire on Sunday night.

For example, this afternoon I heard the familiar tinkling melody of an ice cream truck. People ran toward the truck, waving coupons to be exchanged for the free ice cream sandwiches and rocket pops. I had gotten a coupon earlier today, as thanks for giving directions to some strangers. Not only does the barter system work in Black Rock city, it's inspirational to see people motivated to be generous to others around them . . . with no immediate incentive. Coming home, I passed by the "Mafia camp," decorated with laundered Monopoly money, garlands of grapes, and Catholic saint candles. Unfortunately, I was greeted with some tragic news. The Don was shot last night! The radio station reports that the Calzone family are among the prime suspects. My companions and I on our guard as we show up to the "funeral" dressed in black, with lace veils and a bottle of wine for the grieving "family." When the group swells to over 60 people, the Don is placed on a funeral "barge" (a small motorboat on a trailer) and paraded through center camp. The Calzones attack with water guns, and after a much-appreciated soaking, we retreat to our camp to prepare for the evening ahead. By 8 pm, people are streaming out to the man from all directions of camp. A parade of floats, fire breathing stilt walkers, glowing fairies, and other apparations (*sic*) circles around the man, and then the crowd is pressed back to a safe distance. The burn is spectacular, and I scream and howl until I am hoarse. As the pyrotechnics are lit, the Man is hidden by a brilliant white light; then he appears through the smoke and majestically falls to the ground. We make our way back to the village, where it's now time to burn our individual offerings. As a community, we helped each other create our projects, and now it's fitting that we are together when they are burned. We circle around the "Jenni Board," dedicated to a friend who died this year, now covered with dedications from other participants. A 15 foot blue mushroom is lit on fire, and then we walk over to my project. The Tower takes several tries to light, but once it is on fire, it burns hot and brightly. I solemnly cast a bracelet that I have worn all year, in remembrance of my commitment to this project, into the flames. Now that it's over, I breathe a sigh of relief. I realize that it's not about the size or beauty of the sculpture, it's the experiences in creating, sharing, and destroying that leave the biggest impression. The art at Burning Man may not be the most meaningful, or even made by "real" artists; it's made to be seen, touched, encountered. There doesn't always have to be a point or a meaning. It's about challenging your boundaries and not taking anything too seriously. (http://www.burningman.com/archive)[1]

It is easy to understand why researchers need to provide detailed descriptions of participants' observations because any attempt to interpret a community festival on one simple level can become an analytical nightmare.

Definitions of festival

In the opening article in his edited work, *Time Out of Time: Essays on the Festival*, Falassi defines festival as "an event, a social phenomenon, encountered in virtually all human cultures" (1987, p. 1) which, though true, hardly tells us anything at all. According to folklorist Roger Abrahams, "festivals manufacture their own energies by upsetting things, creating a disturbance "for the fun of it" (p. 178)." Abrahams contends that in western society, we tend to associate ritual with being "for real" and festivals with "fun" (p. 177). However, to the participants of Burning Man, Fasching or Day of the Dead, their festivals are very "real." As a matter of fact, the planning and organization for the following year's festival are often so involved that work on the next festival begins right after cleanup.

Over the years, a number of historical, sociological and psychological approaches have been applied to understanding what constitutes a festival. For example, since Victor Turner introduced the concept of structure and anti-structure in 1969, many anthropologists and folklorists have considered carnivals as prime examples of cultural rites of "inversion." In other words, restoring "order" after festival behavior reaffirms social codes and harmony. "Everyday" routines are resumed and the community waits until the next festival occasion to invert its regular, accepted social rules. Turner believed that the agreement of community participants to celebrate together in a certain manner at an agreed upon time indicated that these community members shared a feeling of *communitas* or belonging to a group. That is all well and good except that within every festival there are clearly distinct "subgroups" such as certain segments or float clubs in the New Orleans' Mardi Gras that are far from "inclusive" (Kinser, 1990). Conversely, anthropologist James Fernandez (1986) contends that the 'meaning' of festivals is to convince not only others outside community but its very members that they have a distinct shared group identity.

Because of the flexible and complex nature of festivals, however, it would be simplistic to suggest that any one of these really explains what constitutes a "festival." Furthermore, most of these theories fail to give credit to the nature of festival play as an enduring statement in itself. The variety of interpretative approaches above are in large part made possible because of the labile and playful nature of festivals like Burning Man, Fasching, and Day of the Dead. Indeed, even malls and businesses have inserted the festival experience into consumer society by adopting the accoutrements of the festival.

Rather than to reduce the complexity of community play to a limited definition of the word 'festival,' I suggest here that we focus on a way to discuss the many facets of festival as 'community at play.' There is a series of eight distinct aspects of festivals that serves as a reliable approach to understand the widespread importance of community play: form, space and time, agencies, roles, transcendent motives, function, organization, and rhetoric. When these elements are made explicit, we can discern the ways that the business world attempts to transform American mall shopping into a 'festival experience.'

FORM

The complexity of a festival is usually determined by the number of events that make up the festival as a whole, and there is often an opening and closing ceremony. Some festivals take the form of a procession, such as the Fourth of July parade. Other festivals do not utilize a linear structure such as the parade and may instead have many events occurring simultaneously. The latter is certainly true at today's American malls where several activities and events are scheduled throughout the mall walkways. The mall festival begins as soon as the shopper enters through the mall doors. At each turn of the corner, the shopper is greeted by another performing group or greeted by a sales associate who attempts to attract the shopper into the store to sample a free squirt of perfume. The experience ends when the shopper exists through the mall doors to return to the "real world," just as one enters and exits any festival grounds.

In a similar way, the African-American festival, Odunde, is both linear and multi-event. This festival "begins" at the Schuylkill River where participants meet the Elders who walk along with the *Egungun* masked dancers and a parade of masked dancers and drummers. Thus, this festival parade begins when enough people assemble in the marketplace perhaps around noon or as late as 1:30 p.m. A parade led by masked Egungun dancers and a Nigerian drum battery march to the South Street Bridge. Here, at the "crossroads" the year's "Elders" bless the colorful flowers and fruits that are tossed into the Schuylkill River. Before the ceremony, honey is poured on the ground at this crossroads while participants dab honey on the backs of others' hands as a gesture of good luck. In a sense, this is a gesture similar to *ebo. Ebo* in Yoruba culture can be an offering or a shield. *Ebo* is best understood as a religious act consisting of ritual procedures for establishing communication with spiritual beings in order to bring luck to persons on whose behalf it is performed. This is a good example of the combination of play with religion since the majority of participants do not practice an African religion. Religion too is a cultural system, and acknowledgement of "syncretism" in African-American cultural expressions reveals the festival as a polymorphous space where elements of an understood African heritage and elements specifically unique to the African-American experience are displayed. The use of African images is also a conduit for festival processes, which at the same time emphasize the importance of Africa as a component of the African-American identity.

The blurring of the sacred versus secular distinction is consistent with anthropologist Margaret Drewal's work on Yoruba celebrations of Nigeria. In Yoruba language, the words play, festival, spectacle and ritual are used interchangeably, and, "When Yoruba people say that they perform ritual 'just like' their ancestors did it in the past, improvisation is implicit in their re-creation or restorations" (Drewal, 1992, p. 23). Improvisation or play among the Yoruba, notes Drewal, is like "signifying" among African-Americans which is a disruption of the signified and signifier that opens up meaning. An example would be when one plays with the frames in language. Signifying both breaks with tradition yet revises the most salient features of the tradition. In a double sense, the Odunde festival is itself an act of signifying: it opens up a profane public space which becomes a place where street religion is celebrated through dialogic festive popular and traditional forms that produce multi-layered messages, thus establishing clearly Bateson's "play frame" (1956).

It is the blurring of boundaries which is an essential element that opens up the space for new playful combinations where, in festivals like Odunde, religious elements are "played" with. More New Year's offerings are tossed in unison as gifts to Oshun, the Yoruba goddess of the river as gestures of *ebo* for the community. After this solemn ritual exchange of reciprocity, Odunde participants return to the street to enjoy a day of music, dance, food and fun along Philadelphia's 21st and 24th streets while the *Egungun* as "shrines in motion" dance among the participants.

This blurring of religious and play boundaries is not unusual. For example, during the Jewish religious ritual of Seder, children are allowed to engage in a game in which the *afikomen* (the matzo eaten at the seder meal during Passover) is taken "ransom" and hidden by the children for a ransom since the Seder cannot end until the *afikomen* is eaten. During the Jewish religious Festival of Lights, Hanukkah, games of chance are played such as dreidel games or card games with a regular deck of cards disguised as a "Maccabean deck." As with most festivals, competition, even in highly religious celebrations such as these, is present, whether overtly as a contest or more subtly as demonstrated here in Seder and Hanukkah. And just as it has become culturally appropriate for Jewish families to purchase Hanukkah greeting cards and Hanukkah wrapping paper to wrap the Hanukkah gifts that they've purchased at American malls, the blurring of religion and commerce certainly occurs as well during the Christmas season in the mall.

SPACE/TIME

American businesses are often housed in malls which are enclosed spaces, set off from the "frame" of daily life by their separate locations from residential and "typical" business areas. This very act of a separate space, as in all festivals, signals to the community this space I different from other spaces in our everyday life. In festivals like Odunde, the introduction of elaborate African images makes the event more spectacular and helps to localize the festival. An Odunde banner becomes a point of reference for the festival and, as the crowd moves into this area, paradoxically the space is "opened up" for festival, thus creating the center of the festival in a world of highly coded festival vocabulary. In southwestern Germany during Fasching, a clothesline stretched across a town street from one apartment window to another and laden with underwear such as different colored bras takes the place of flags or banners. In some of the German towns a "carnival tree" is decorated with unusual decorations such as shoes or brooms and erected on a high town steeple. In all cases, a closely packed crowd is desirable because this creates a feeling of oxygen deprivation, and the density of people contributes to intensification of the festival.

When such internal tensions are created, there is a feeling of communitas among the participants. Filling the air with sound is another way that festivals operate to dislocate and relocate participants. In the Odunde festival, the use of both African and hop-hop cultures blurs boundaries, thus bringing people together.

In American consumer society, the business community has recognized the power of the festival experience and its attraction for many people. The business world often attempts to create this dislocation of time and space for consumers. For example, when entering the American mall, shoppers are seemingly transported to a distant "time out of time," (Victor Turner's liminal time) with banners and brightly colored corridor decoration to signal that the shopper is entering another reality where the same rules do not apply. The benefit of creating a festival atmosphere for shoppers is apparent: festival is a time for release from the everyday world; in the festival world, there is a sense of "freedom" and one is encouraged to eat and drink whatever is appealing at the moment. This encouragement to ignore usual limits takes the form of overspending for many and may explain why many Americans of all ages and gender find the mall exhilarating until credit card invoices start arriving the following month.

AGENCIES

Certain festival codes or agencies are a part of every festival experience, whether it is a mall, county fair, or parade. Costumes, masks and clowns are often agents that are introduced during festivals to create a feeling of festive "vertigo." Flags and banners are often displayed high above shoppers in American malls and sales employees costumed as Big Bird or Santa roam casually about the walkways. These figures may take the form of Uncle Sam on stilts during the Fourth of July or clumsy clowns like those found at county fairs.

At an ethnic festival like Odunde, it is easy to see why the African images are central to the success of the festival. The stranger or "it" figure in festivals (as in "Who is it?") activates the crowd by intensifying the festival experience: masked strangers, who are really not strangers but rather disguised members known to the African-American community, are introduced into the festival just before the parade.

These masked *Egungun* twirl as other masked deities walk on high stilts. Stilted figures stride above the crowd as a way of elevating the gaze of the crowd so that the normal line of sight is upset. This figure galvanizes the attention of the group because it appears very strange while twirling masked figures help to sustain this sense of vertigo from below. Throughout the festival day, the stilted figures continue to activate the crowd by strolling throughout the crowded streets, sometimes appearing to sway precariously. The festival mall figures costumed as Santa or the Giant Pumpkin waving to shoppers and patting children on the head help to get shoppers in the mood to participate in this consumer's paradise.

In most festivals, special foods are agencies that inform participants that this is a special time, a "time out of time." In addition to spontaneous and scheduled staged musical performances, another major element of Odunde is the African market, which sells traditional African-American foods such as sweet potato pie along with foods from the Caribbean. The combination of both American foods and traditional African foods signifies that this is a special time for that community. In the modern mall festival, the food court often offers an array of international and deli foods in addition to elaborate ice cream flavors and coffee concoctions that one may not normally consume at home.

ROLES

During festivals, images are often larger than life or situations are reversed. Roles are often inverted or exaggerated. During Fasching, the German carnival, a masked figure will replace the town mayor. In Odunde, the Elders of the festival are honored. Uniforms and costumes or face paint enable particular groups and persons to communicate their festival positions with non-verbal behavior. Much of this communication is coded in what is called "festival vocabulary" which can be easily decoded by community members. During sporting events, the players are distinctly marked but spectators also communicate their affiliation with teams by the use of color, costumes or face paint. Team mascots who are the equivalent of festival clowns and masked "it" figures that represent outsiders heighten the festival experience through comic behaviors or special performances.

During community play, everyone can take part and thus everyone has a role, even if it is one of spectator. The relationship between the team mascots or cheerleaders and clowns and the spectators at a festival or sporting event is dependent on call and response activities and eye contact as well as the use of distinctive dress. The role of the costumed participants is to engage the spectators but at the same time to keep them at a distance. At the mall, shoppers are encouraged to feel that they can be part of the "action" when they purchase goods in the stores. Mimi-amusement rides in the mall also encourage active participation by shoppers en route from one store to another. On November 23, 2001 (also known as "Black Friday" because most American stores go from being in the red to balancing their books, thus being in the "black"), the Bon Ton department stores in the malls in Syracuse, NY had salesclerks greeting customers at their entrances with a "purchase and win" game card. The game card resembled bingo and thus, early bird shoppers had "chance" to "win" discounts, that could be "scratched" to reveal the amount only AFTER the purchases were in the process of being made, thus introducing a bit of game competition into the mall "festival" experience.

TRANSCENDENT MOTIVES

Metaphor plays a big role in much of festival performance which forces commonly accepted values of the community through exaggeration in highly coded events. For example, competitive sports and even county fair tractor pulls celebrate aggression but in a highly coded way. In American society, competition is a fundamental value in a marketplace culture and plays a central role in American football events, pie contests at County Fairs and Buck Contests during deer hunting season. Often highly individualized sports like deer hunting or ice fishing is transformed into a community festival by the introduction of competition. The Blue ribbon awarded at the County Fair for prize-canned goods and large vegetables may be interpreted as both competition in the marketplace or a symbol of productivity but often the same vegetable may be appreciated for its comical or bawdy shape. Here we recognize the labile aspect of play: what seems to support a value system can at the same time subvert it.

Does metaphor play a role in the mall festival experience? Certainly, if we consider that the capitalist system itself is supposedly based on competition and free enterprise. At the mall, stores compete to offer the best, the most for the least amount of money. Shoppers are encouraged to "comparative shop" and "early bird specials" crop up at different stores. Shoppers waiting in lines compare prices and quality of the merchandise that they have purchased. And after the Sept. 11th tragedy in New York City, it suddenly became "patriotic" to shop at the mall. Almost every American store throughout the country quickly displayed the American flag. From a bird's eye view, we might say that the mall itself is a microcosm of what is valued in American culture.

FUNCTION

Festivals provide a time and space to celebrate past events or to envision alternatives. Festivals may subvert or even challenge societal norms through competitive events. By their very nature, festivals can become explosive. Consider the aftermath of many sporting events when males engage in fighting or even skating performances that seem orderly but instead subject the dancer to possibility of injury. For example, dancing on roller skates is unpredictable and may even be dangerous:

> Ideally, then the skating can enhance the skater's sense of self and of belonging. In this vein, the skaters' games, like some children's games, has (sic) potential for personal development and creation of community. Ideally, again, it could offer a better future to those in a difficult present by teaching adaptation and assimilation skills to the players. However, like children's games, adult play is not always safe, innocent, and beautiful. Moreover, some adult play can be more self-conscious than children's play, purposefully combining physical, even sexual and symbolic elements with social as well as solitary activity for a maximally vertiginous effect.

This kind of play toys with the danger and manipulation inherent to the exhibitionism of performance and spectacle (McGovern, 1993, p. 91).

Consider as well, the following journal entry from one festival-goer at the Burning Man festival:

> The heat is also causing tensions in our camp to run high. We are all frustrated because there's so much to see and do, and not enough time or motivation to leave the village for very long. Several times we've barely managed to start up the stove, and strangers have appeared at our impromptu kitchen, demanding free handouts of food. Even though our fresh food supply is low, sharing isn't the problem. It's the attitude of these people that is really surprising . . . they show up unprepared and then expect that others will do all the work and clean up afterwards. It's definitely a window into the ugly side of human behavior, and an unfortunate result of a quickly growing community. (http:www.burningman.com/archive)[1]

This participant's recollection is very different from Turner's communitas and the claims about the wonderful camaraderie at Burning Man. The contradiction is apparent when we analyze what is called "festival rhetoric." Elsewhere, I have written about the ways in which deer hunters engage in community play such as hunting and divide themselves into "us-them" groups. This "us-them" attitude creates a stronger bond among the local deer hunters but separates the locals from the "city hunters" who come to the Adirondack Mountain region in upstate New York to hunt every year (McMahon, 1993a, p. 154). Thus, participation in the same sport within the same region and at the same time does not guarantee that unity among the entire group is created during community play. In the Adirondack hunting case, if anything, it reinforces the "outsider-insider" conflicts that exist in the region year round. This may be so about the American shopping world but paradoxically, the mall brings together Americans from all ethnic groups and economic classes. It some ways, it is an attempt to create a utopia or shoppers paradise.

ORGANIZATION

One of the first questions an observer of any festival should ask is who organized this event? Festival participants often refer to their festivals as spontaneous. Yet, festivals are most often scheduled according to the calendar. Often permission must be secured in advance because, for example, in urban environments zoning ordinances may forbid public events from occurring in a certain locale. Consider the plans below for the First Night Celebration, an annual festival held in a small town Oneonta, New York on December 31. The Steering Committee for this small festival consists of nine members including a Chair, Vice-Chair, and other appropriate officials who head the following volunteer sub-committees who carry out the following listed duties:

Logistics:
  arrange for the safety and security of the participants through
      agreement with OFD (Oneonta Fire Department), Red Cross, OPD
      (Oneonta Police Department) and/or private security companies se-
      cure all approvals required by City Government, as well as insurance
      and other licenses as needed.
  arrange with OPT (Oneonta Public Transportation) for
      transportation, and determine bus routes
  arrange for the support of other City Services (i.e.
      DPW)(Department of Public Works).

Venues:
  determine available sites
  determine and address the needs of each site
  oversee decoration
  assist the house managers as needed
  arrange for set-up and clean-up
  determine availability of food and facilitate use of vendors (where
      appropriate)

Fireworks:
  contact and contract with Fireworks provider
  contact FAA for approval

Volunteers:
  recruit and train volunteers from businesses, service groups, and
      churches, as well as individuals
  determine volunteer assignments and schedules
  assist the house managers as needed

Finance:
  perform the functions of Treasurer
  determine and monitor budget
  fund-raise through telephone, mail, and personal contact

Marketing/Public relations:
  write, design, produce (and distribute as appropriate) ads, fliers,
      brochures, programs, posters, and signs
  supply logo info and approve banners
  write and disseminate press releases
  write & coordinate placement of radio and print ads
  make presentations to schools and organizations
  conduct interviews
  update and maintain web site
  determine and coordinate in-school promotion
  facilitate promotion through churches and service organizations

Entertainment:
   solicit new entertainment, send out applications, coordinate a
   review process and select entertainers.
   working with pre-determined budget, negotiate contracts with
   performers
   schedule performances
   address performers needs through coordination with venues and volunteer
   committees
   coordinate parade
   arrange for transportation of puppets
   arrange for puppet making workshops

(NOTE: Puppets will need to be moved from outdoor to indoor storage at Lettis
Auction House when snow has melted. In addition, Catskill Puppet Theatre will
need to be contracted to repair small puppets.)

Buttons:
   determine design of button and coordinate production
   arrange for points-of-sale and distribute buttons
   establish a system of control and record keeping
   collect and catalogue button sales money

5K Race:
   solicit sponsorship
   design tee-shirt and coordinate distribution
   determine race course and (with Logistics Committee) enlist City
      Government support
   prepare and distribute race materials
   coordinate the application process
   oversee race and awards ceremony
                    (Xerox of the Steering Committee, 1999)

   This First Night celebration is a far cry from the complexity of large festivals which span more than one day, such as Mardi Gras in New Orleans. Yet,
even to create this small festival, the committee in Oneonta meets in February to
begin plans for the next December 31st festival.
   On a much larger scale, many of the same concerns are currently being addressed in Syracuse, New York where I reside. With the process expansion of
the Carousel Mall, City Council members and other community residents are in
the process of working out what is to be allowed, what should be included in the
plan to make certain additions that the community desires and what is necessary
for this expansion to meeting legal regulations. No matter how large the festival
however, the organizer of a festival is known to local community members and,
if the festival is extremely large, one can still easily find out who initiated the
festival by enquiry.
   For example, in June 1992 I first encountered Odunde, an African-
American festival in Philadelphia. The Odunde festival was first organized in
1975 by community activists Lois Fernandez and Ruth Arthur who applied for a

$100 grant from South Central Philadelphia Affairs Congress (PAC) and created a committee composed of the women of South West Center City Citizens Council. Lois Fernandez is recognized by Odunde participants are the originator of the festival. Like Lois Fernandez, in 1985 with little money and a simple eight-foot structure that burned in impromptu fashion, Larry Harvey started the Burning Man festival. The Burning Man festival has had to move three times from private land to Desert Rock. Both Lois and Larry had to overcome regulations and local ordinances which required their leadership, time, effort and organization. Here is how Larry Harvey, in an interview (Hualapai Playa, Nevada, August 30, 1997) describes the process which is far from spontaneous:

> Now our story as planners, we plan and plan and plan so this spontaneous thing that you do will happen. And our experience has been pretty . . . has been tough this year. Because planners . . . as planners we've had to look at our economic realities. There's boundless wealth here, look at . . . everybody's giving everything to anybody else, I've never seen anything like Boundless wealth. But in order to do the planning, of course we need money. Of course we money Now, when we came in to the county, we talked to the service . . . we talked to the agencies and the fire folks and so forth, and I have no complaints of them. None. They . . . they're in the business as us. If you organize something this wild you'd better be responsible. You'd better think about people every minute of the day. They're people who've worked in the field and they know they're talking about and we learned a lot from them. The only problem is something interrupted that process. You know, our first conversations with them, we were on the same wavelength. And then the politicians got involved. Now a politician is someone who spends all of their time . . . trying to avoid being blamed for anything and trying to take credit for everything else. That's what they do for a living. So there was what we call a caucus. And after that caucus things started to change. Suddenly . . . suddenly the reasonable arrangements we had made, out of mutual concern for public welfare . . . oh, those discussions . . . weren't possible now. Now we're talking by the book and literal. Now we're talking about . . . suddenly as we went on in this process we were talking about, we were talking about mandated services that we couldn't bid out. That we got lower . . . way lower bids for. We couldn't bid out. And then we were talking about . . . well, it goes on and on. I won't burden you with it. Have you ever dealt w i t h   a   b u r e a u c r a c y,   y o u   k n o w. (http://www.burningman.com/project/history/hualapai2.html)[1]

No festival continues without bureaucratic support of some kind such as city officials, church groups or social welfare groups. More importantly, there is usually the required cooperation from police. Even during the Fasching celebrations in Germany when carnival clowns "take over" city hall and the mayors in towns and cities literally hand the city keys over to them, there are unspoken regulations or acceptance of a "disorderly order" which keeps the carnival under control. As Kinser (1990) maintains, even so-called 'subversive festivals,' such as carnival, must conform to official norms: "'Officially' Carnival is an 'unofficial activity,' a private affair, but it is clear to the city's politicians as it is to most citizens that Mardi Gras is an unbelievably good commercial investment" (p. 294).

Thus, within the festival organization itself, there is another kind of bureaucracy: the organizing committee has to develop a plan which will be acceptable to all. There are usually many subcommittees that are formed to oversee individual events, and this process is repeated ad infinitum depending on the complexity of the festival. However, spontaneous play sometimes occurs within the larger planned community play. These forms of play can take the simple form of a joke or an impromptu prank played by festival goers. Some of these impromptu pranks became traditional events of subgroups who attend the larger festival. Consider the following description by one participant at Burning Man:

> Tired as we are, we never miss a chance for playful interaction. Dressing in cheap Santa suits is a several-year tradition, and last night, over 30 of us met to bring a little Cacophonous Christmas cheer to Black Rock City. We responded to idle threats from the clown camp, and "raided" their circus ring. Red rubber noses were tossed into the air with Santa beards during the playful scuffle. We ended up at a local dive in the "NeighBARhood," laughing   with   some   disgruntled   postal   workers. (http://www.burningman.com.archive)[1]

RHETORIC

What are the beliefs, attitudes, and feelings behind the festival? How does one discern the "meaning" of community events? And what are the rhetorics of looking at the American mall experience as a kind of community play? Fourth of July parades are said to be tributes to the Land of the Free and patriotism. A community's Mother's Day Dinner is said to be held in honor of motherhood; the county fair represents the community's productivity or the region as a "land of plenty" In addition to these popular rhetorics, unique to each individual community's festival, Sutton-Smith points to the ideological rhetorics of scholars from various disciplines who study the festival. In his identification of these scholarly play rhetorics; Sutton-Smith (1999) locates seven criteria for identifying each:

1. That each rhetoric is a thesis about play's *function*;
2. That the rhetorics can be shown to have a clear base in some general *cultural attitudes* of a contemporary or historical kind;
3. That each rhetoric has its own distinctive group of scholarly *advocates*;
4. That each rhetoric applies to a distinctive *kind of play*;
5. That each rhetoric applies to *a distinctive class of players*;
6. That the group of players, their organizers, their supporters and rhetoricians often *establish forms of hierarchy* and hegemony internally over their players and externally over their competitors or over those who are members of other playforms;

7. And, finally, there are variable *affinities* between the rhetoric and the scholarly discipline which subsumes it, and between the rhetoric and the type of play to which it refers (1999, p.151). Using these criteria, Sutton-Smith suggests that in the festival literature itself, there exist three modern and four ancient rhetorics: in the modern world, scholars use progress, imagination, and the self to explain the phenomenon of festival. These rhetorics are predated by earlier scholars who claim that power; identity, fate and frivolity explain its existence. Progress rhetoric focuses on play as adaptive behavior, both biological and psychological; the rhetorics of fate translate into an existential optimism; power in play is seen as contest or hegemony; identity rhetorics promote play as social context or *communitas*, Turner's term for "group flow;" rhetoricians of the imaginary claim play is transformation; rhetorics of the self focus on play as a peak experience, Csikszentmihalyi's "flow," and frivolity rhetoric promotes the idea that play is a rite of symbolic inversion. However, Sutton-Smith demonstrates that no single theory is sufficient and further, "tells us no more about play than the use of applied music in classrooms tells us about the theoretical fundamentals of music" (1999, p. 152). On play in general, Sutton-Smith (1997) shows us that all we really know about play is that it is exciting for children, adults and even many animals. What play IS still eludes us. This leaves it fairly open for researchers to ask, is mall shopping an American form of community play?

CONCLUDING REMARKS

Sutton-Smith (1997) like others notes that it is simplistic to see the notion of freedom in play as non-productive. Abrahams elucidates this idea: "But surely more is meant by freedom than simply the relief from the need to be productive in expenditure of energy. Specifically, especially in emporium cultures, the act of buying and selling products exempted from the rules of family and community exchange is itself liberating" (1990, n. p.). Abrahams' observation in and of itself gives us reason to consider the contention that the American mall provides a festive experience for many Americans. Coupled with the individuals' comments about their enjoyment of shopping at the mall with hundreds of other shoppers and that after the Sept. 11th terrorist attack in NYC store signs promoted shopping as a "patriotic" activity, it does not appear to be a far stretch to consider that mall businesses are intentionally striving to provide a festival experience for customers.

By looking at both festival vocabulary and the underlining rhetorics of play, theorists may provide a better framework with which to understand the processes at work during community play. However, researchers must proceed with caution before making any generalizations about meanings for individual participants and performers because every time the festival is "played," it is slightly different than the festival before. For each performer and festival-goer, histories change, memories fade. Yet it may be possible to learn more about the ways in which meanings are generated by focusing on the vocabulary of the festival over time and, more specifically, by focusing on the organization and organizers of the festival each year. As I have written elsewhere (McMahon, 1993a), traditions such as festivals and other forms of community play are symbolic expressions created through a negotiation process that must result in some kind of community consensus as to what will be "played." Since differences among the group's organizers have to be worked out before there is a consensus, we should be looking at these differences. Differences do not necessarily cease to exist when a seeming consensus is reached. The same differences may surface later when tensions within the group or when conflict with those outside the group arise. It is through re-negotiation among the group organizers that the symbolic expressions may change in this continual re-negotiation of festival images. This process is ever in flux and may account for the academic neglect of looking at the American mall as a traditionally appropriate festival site within American value system. "Money still has the right to wear a costume and a mask in this never-ending carnival, and referendums have proven that the majority of the (American) population finds nothing wrong with that" (p. 136). In spite of Galeano's intended cynicism, with regard to the American mall, taken literally his observation experience seems quite accurate.

NOTE

1. Disclaimer: As is the nature of the world wide web, websites in this chapter which were accessed in 2000 may no longer exist.

REFERENCES

Abrahams, R. D. (1987). An American vocabulary of celebrations. In A. Falassi (Ed.), *Time out of time: Essays on the festival* (pp. 173-183). Albuquerque: University of New Mexico Press.

Abrahams, R. D. (1990). *The marketplace experience and festive play*. Unpublished manuscript.

Babcock, B. (Ed.). (1979). *The reversible world*. NY: New York University Press.

Bakhtin, M. M. (1981). The dialogic imagination. Austin: University of Texas Press.

Bateson, G. (1956). The message "This is play." In B. Schaffner (Ed.), *Group process* (pp. 145-242). NY: Josiah Macy.

Bauman, R. & Abrahams, R. D. (1978). Ranges of festival behavior. In B. Babcock (Ed.), *The reversible world: Symbolic inversion in art and society* (pp. 193-203). Ithaca, NY: Cornell University Press.

Carmichael, E. & Sayer, C. (1991). *The skeleton at the feast*. Austin: University of Texas at Austin Press.

Csikszentmihalyi. M. (1990). *Flow: The psychology of optimal experience*. NY: Harper.

Drewal, M. T. (1992). *Yoruba ritual: Performers, play, agency.* Bloomington: Indiana University Press.

Durkheim, E. (1915). *The elementary form of religious life.* London: Allen and Unwin.

Falassi, A. (1987). *Time out of time: Essays on the festival.* Albuquerque: University of New Mexico Press.

Fernandez, J. (1986). *Persuasions and performances: The play of tropes in culture.* Bloomington: Indiana University Press.

Fernandez, J. (Ed.) (1991). *Beyond metaphor: The theory of tropes in anthropology.* Stanford, CA: Stanford University Press.

Galeano, E. (2000). *Upside-down: A primer for the looking glass world.* NY: Holt & Co.

Geertz, C. (1972). *Deep play: Notes on the Balinese cockfight.* Daedalus 101:1-38.

Gluckman, M. (1963). *Order and rebellion in tribal Africa.* Glencoe, IL: Free Press.

Kinser, S. (1990). *Carnival American style.* Chicago, IL: University of Chicago Press.

Leech, E. (1961). *Rethinking anthropology.* London: Athlone.

Manning, F. (1983). *The celebration of society: Perspectives on contemporary cultural performance.* Bowling Green, OH: Bowling Green University Press.

McGoven, J. (1993). Reconsidering rollerskating: The body moving between play and performance. *New York Folklore, XIX* (3-4) 87-97.

McMahon, F. (1993a). Regional sports: "Playing" with politics in the Adirondacks. *New York Folklore, XIX* (3) 4:59-73.

McMahon, F. (1993b). The worst piece of "tale:" Flaunted "hidden transcripts" in women's play. *Play Theory & Research, 1*(4) 251-258.

McMahon. F. (2000). The aesthetics of play in reunified Germany's carnival. *Journal of American Folklore, 113*:450 (Fall) 378-390.

Rabelais, F. (1965/1532). *Gargantua & pantagruel.* Baltimore: Penguin Books.

Schechner, R. (1988). Playing. *Play and Culture, 1*(1):3-27.

Stally-Brass, P. & White, A. (1986). *The politics and poetics of transgression.* Ithaca: Cornell University Press.

Sutton-Smith, B. (1995). Conclusion: The persuasive rhetorics of play. In Anthony D. Pellegrini (Ed.), *The future of play theory: A multidisciplinary inquiry into the contributions of Brian Sutton-Smith.* (pp. 275-296). State University of New York: Albany Press.

Sutton-Smith, B. (1997). *The ambiguity of play.* Cambridge: Harvard University Press.

Sutton-Smith, B. (1999). The rhetorics of adult and child play theories. *Advances in Early Education and Day Care, 10*:149-161.

Swiderski, R. M. (1986). *Voices: An anthropologist's dialogue with an Italian-American festival.* Bowling Green, OH: Bowling Green University Press.

Turner, V. (1969). *Ritual process.* Chicago, IL: Aldine.

Zepp, I. G. Jr. (1986). *The New Religious Image of Urban America: The Shopping Mall as Ceremonial Center*. Westminster, MD: Christian Classics.

Chapter 13

Validity and Reliability of the Experience of Leisure Scale (TELS)

Cheryl R. H. Meakins, Anita C. Bundy, and Jeffrey Gliner

INTRODUCTION

> If we are to realize the full range and potential of recreative leisure, we shall have to be perceptive enough to think of it as something more vital.
>
> (Brightbill, 1960, p. 11)

Leisure and play as buffers to stress and illness in adults have become topics of great interest (Cohen, 1993). For example, Des Camp and Thomas (1993) found that physical play decreased nurses' stress while Cunningham (1989) reported an inverse relationship between stress and satisfaction with leisure. Finally, Shaffer, Duszynski, and Thomas (1982) found an association between intellectual entertainment (e.g., books, concerts, and art) and decreased frequency of cancer in physicians. In addition to promoting health and reducing stress, Glynn (1994) found that play increased productivity.

Leisure and play are not synonymous. While leisure seems to be a larger construct than play, they share certain overlapping traits. For example, most theorists seem to agree that both play and leisure occur in the context of freely chosen activities. Like play (Neumann, 1971), leisure can be defined as an *experience* (rather than as discretionary time or specific activity) in which the player is relatively intrinsically motivated, internally controlled, and free from some constraints of reality (Blatner & Blatner, 1998; Brightbill, 1960; Kerr & Apter, 1991; Neulinger, 1974; Schaefer, 1993; Witt & Ellis, 1989). One difficulty with such a definition is that it might allow for the inclusion of activities that are potentially harmful to the individual (e.g., drinking, drug use), society, or the environment (Csikszentmihalyi, & Kleiber, 1991; Mannell, 1993). Potentially harmful activities are not ones that helping professionals seek to promote with their clients. In contrast, desirable leisure activities embody these traits *and* lead to self-actualization. Such activities also have been called committed or serious leisure (Csikszentmihalyi & Kleiber, 1991; Hogg, 1993). Serious leisure typically is marked by intense involvement of the participant (Csikszentmihalyi & Kleiber, 1991; Hogg, 1993). In the assessment described herein, we have adopted a model of leisure based on two assumptions: (a) activities in which individuals feel relative intrinsic motivation, relative internal control, and freedom from some constraints of reality are likely to be experienced as serious leisure; and (b) serious leisure activities in which one becomes totally involved are likely to contribute to self-actualization.

While leisure is important to health and adaptation, few tools exist to measure it. Measurement of leisure may be particularly useful to helping professionals—therapists, social workers, psychologists, and other professionals—whose focus is on interventions directed toward quality of life. Helping professionals commonly use checklists (e.g., The Interest Checklist [Matsutsuyu, 1969]) to allow individuals to identify their preferences for leisure activities. The problem with such tools is that they can be difficult to interpret (Bundy, 1993). Further, since checklists provide a specified set of leisure activities, they do not account for activities that are not socially sanctioned as leisure but that individuals might nonetheless enjoy in that way (e.g., cooking).

The Leisure Diagnostic Battery (LDB) (Witt & Ellis, 1989), another tool commonly used for measuring leisure in adults, was developed in the field of recreation and leisure studies to define individuals' experiences with leisure. The LDB assumes that if certain characteristics are present and conditions met, individuals will receive maximal benefits from recreation. The items of the LDB operationally define the characteristics and conditions of leisure (i.e., perceived competence, perceived control, needs, depth of involvement, and playfulness). The LDB has both short and long forms. The long form yields a collective score that describes an individual's perceived freedom in leisure. Though the LDB can be applied to individuals who have minor physical limitations, it appears likely that those with more significant limitations would be artificially penalized since one major focus of the adult long form is on sports and competitive games. Further, the adult long form has not been tested for validity and reliability.

The LDB short form for adults has been examined with individuals with and without disabilities for evidence of internal reliability (.88 to .94) and validity. In one study of church members, the LDB short form for adults showed a correlation of .27 to the Rosenberg Life Satisfaction Scale (Hensley & Roberts, 1976). Another study found the LDB to have an inverse relationship to the Barriers to Leisure Opportunities Scale (Witt & Ellis, 1989) when examined with members of the National Stutterers Project.

While the short form of the LDB appears to meet many of the needs of helping professionals, it might be difficult to use it as the basis for intervention planning. The LDB does not establish a context in which individuals assess their leisure; rather respondents are encouraged to think of their leisure overall. Given the wide variety of leisure activities, tools that provide information specific to an individual's experience in a particular valued activity should form the basis for the most effective intervention.

The Test of Playfulness (ToP) was developed by an occupational therapist (Bundy, 2000) to examine playfulness in children with and without disabilities. The ToP was originally designed as an observational assessment and reflects four elements of play: intrinsic motivation, internal control, freedom from some constraints of reality, and framing (i.e., giving and reading cues). Reliability and validity of the ToP have been examined with typically developing children of varying ethnic backgrounds and ages (e.g., Bundy, Nelson, Metzger & Bingaman, 2001; Griffith, 2000; Porter & Bundy, 2000; Previd, 1999; Tyler, 1996) and with children who have disabilities (e.g., cerebral palsy, autism, ADHD, developmental delay) (Gaik & Rigby, 1994; Harkness & Bundy, 2001; Leipold & Bundy, 2000).

While the ToP has potential for use in intervention planning, it is an observational assessment. Observation of adults engaging in serious leisure activity presents a number of difficulties. Not all attributes of the experience are easily identifiable, especially when the activity is done alone. Not all activities that individuals choose as leisure are appropriate for observation. Adults often are self conscious about being observed. Thus, in this study, we converted the ToP to a self-report instrument. When the scale was converted from observation (in which attributes are assigned to an individual) to self-report, it became one of experience rather than approach (playfulness). Given our interest in serious leisure (Csikszentmihalyi & Kleiber, 1991; Hogg, 1993), which seemed to share many common traits with playfulness, we named this new scale The Experience of Leisure Scale (TELS).

When taking the TELS, respondents first named an activity in which they become totally involved. This provided a specific context for the 23 questions that followed. The 23 items reflect the 4 elements of the ToP (i.e., intrinsic motivation, internal control, freedom from some constraints of reality, and framing). Modifications were also made to wording of the items so that they reflected adult experiences rather than children's observable behaviors. (See Appendix A for TELS Protocol.)

Because the constructs represented in the TELS have been supported as characteristics of adults' play and leisure (Blatner & Blatner, 1998; Brightbill, 1960; Kerr & Apter, 1991; Neulinger, 1974; Schaefer, 1993; Witt & Ellis, 1989) and because it minimizes some of the problems previously noted in other tools, the TELS may be a good fit for helping professionals. Unlike the LDB, the TELS measures extent, intensity, and skill of playful behaviors, allowing respondents to describe their leisure experience by how often, how much, and how well. Further, the predecessor to the TELS, the ToP, has been used with children who have significant disabilities, suggesting that the TELS also may be shown to be valid for use with adults who have significant limitations.

The purpose of this study was to examine the reliability and validity of the TELS. We addressed the following:

What is the evidence supporting construct validity of the TELS?

Do data from 95% of the items and respondents conform to the expectations of the Rasch measurement model?

Do 95% of responses conform to the Rasch model?
Do the items form a logical progression from hardest to easiest?
How many levels of involvement in leisure are represented by TELS items?
Are there statistically significant differences between TELS scores (leisure experience) of individuals who rate themselves differently for playfulness (approach)?
What is the evidence supporting reliability of the TELS?
Are item error estimates acceptable (< .25)?
How well are levels of involvement in leisure among respondents distinguished by the TELS?

METHOD

Respondents

Surveys were distributed to 202 potential respondents; 193 (95.5%) were returned. Respondents were adults, ages 19 to 66 (M = 26.1) (125 females, 66 males, 2 not given). Data from a total of 184 respondents were included in the analysis. Data from 6 respondents were dropped because the content of their responses suggested they did not complete it seriously; an additional 3 respondents failed to fill out the form correctly. All respondents were selected from a convenience sample of faculty and students of the Departments of Occupational Therapy and Manufacturing Technology and Construction Management at Colorado State University in Fort Collins, Colorado. No compensation was given to encourage participation.

Instrument

The TELS is a self-report assessment of adults' experiences in committed leisure activities. Qualitative and demographic information also are obtained: respondents' age, gender, number of siblings, birth order, number of children, children's ages, and perception of their own playfulness as these may be related to leisure experience. Respondents are asked to think of an activity in which they become totally involved and keep it in mind while filling out the survey. In this pilot study, we allowed participants to select any activity in which they became totally involved, even if that activity typically might be thought of as work (e.g., writing, reading professional literature) or self care (e.g., bubble bath, cooking). If the activity was done alone, then respondents were asked to skip questions pertaining to social interactions.

Adults rated themselves on a 4-point scale reflecting extent (proportion of time), intensity (degree), and skill (ease) on 23 items. For some questions a "not applicable" option also was provided. Five questions pertained to intrinsic motivation, 9 to internal control, 6 to freedom from some constraints of reality, and 3 to framing. Content was kept as parallel to the ToP as possible.

Procedure

Prior to the investigation, a pilot study was conducted with a sample of 19 occupational therapy professional graduate students enrolled in a research methods class. We altered the survey to reflect their feedback. Surveys were distributed to students in conjunction with a university class. At the beginning or end of a preceding class period, the principal investigator explained the purpose of the study and when and where the students would receive the survey. The surveys were distributed after a quiz or exam so students could use time left over from class for its completion. The first page of the survey explained the purpose of the study and directions on how to complete the forms. Completed and uncompleted surveys were returned to a box located near the exit of the classroom to ensure anonymity of respondents. Faculty members in the Occupational Therapy Department were given the survey in their departmental mailboxes. They were asked to return the completed or uncompleted surveys to the mailbox of the researcher.

Data Analysis

Data were analyzed using BIGSTEPS (Linacre, 1991-1994), a computer program for Rasch analysis. Rasch analysis is based on two assumptions: (a) a more capable person (in this case, one who experiences more intense involvement in leisure) has a better chance for success with any item, and (b) an easier item is more likely to be passed by any person. When data from items and persons conform to the assumptions, they are said to "fit the model." Fit to the model is determined by fit statistics. Two types of fit statistics are calculated: infit and outfit. Each is represented by both mean square (*MnSq*) and standardized residual (z) statistics. The desired value for *MnSq* statistics is 1.0; the acceptable range is .6 to 1.4. The desired value for standardized residuals is 0.0; the acceptable range is -2 to $+2$. When both *MnSq* and z statistics for either infit or outfit are out of range, the data in question fail to conform to the expectations of the Rasch model.

Rasch analysis converts raw data into equal interval measures expressed in logarithmic odds probability units (logits). Item difficulty and participant ability measures are placed on the same scale. As applied to the TELS, low, negative logit values indicate easier items and respondents less involved in their leisure. High, positive logit values indicate harder items and respondents more involved in their leisure. The mean score is 0.0.

RESULTS

To examine validity of the instrument, we checked fit of data from the items and respondents to the Rasch model, logic of the sequence of items, and separation of items into levels. We also tested for statistically significant differences in

TELS scores for groups of respondents who categorized their playfulness on a 0 to 3 scale.

Data from 91.4% of items conformed to the expectations of the Rasch measurement model as defined above. This was slightly less than the desired 95%. The items with data that failed to fit are "engages in simultaneous activities" and "social role." (See Table 13.1.) Fit statistics for "engages in simultaneous activities" were higher than acceptable indicating data were erratic, whereas those for "social role" were lower than desired, suggesting either lack of variability in the data or data that were too perfect. Thus, we examined both these items further. We will discuss them below.

Data from 88.6% of respondents (*n* = 163) conformed to the expectations of the Rasch modelalso less than the desired 95%. Twenty-one respondents' data failed to fit the model.

Table 13.1
Items of the TELS in Measure Order, From Hardest to Easiest

Item	Measure	Error	Infit		Outfit	
			MSq	STD	MSq	STD
Pretends Extent	2.58	.11	1.18	1.5	1.17	1.3
Humor	1.37	.10	.91	-.9	.89	-1.1
Imaginative Use Of Objects	1.25	.09	.83	-2.1	.82	-2.2
Exuberance	.95	.09	.96	-.5	.95	-.6
Pretends Skill	.93	.12	.99	.0	.97	-.2
Teasing/Mischief Extent	.80	.09	1.11	1.2	1.10	1.1
Gives Cues	.48	.10	.69	-3.4	.68	-3.4
Social Role[a]	.09	.11	.56	-4.8	.57	-4.5
Shares	-.17	.11	.99	-.1	.93	-.6
Supports	-.18	.11	.60	-4.1	.60	-3.9
Social Engagement	-.19	.11	1.25	2.0	1.20	1.6

Continued on next page

Table 13.1 – *Continued*

Item	Measure	Error	Infit		Outfit	
			MSq	STD	MSq	STD
Persistence	-.22	.10	.83	-1.7	.92	-.8
Enters	-.26	.12	.97	-.2	1.01	.1
Negotiates	-.28	.12	.91	-.8	.89	-1.0
Teasing Mischief Skill	-.34	.11	1.04	.4	1.05	.4
Process	-.46	.11	1.26	2.3	1.19	1.6
Modifies	-.47	.11	1.22	1.9	1.28	2.3
Decides	-.66	.11	1.22	1.9	1.24	1.9
Initiates	-.67	.13	.82	-1.5	.81	-1.5
Feels Safe	-.83	.12	1.12	1.0	1.28	2.0
Simultaneous Activity[a]	-1.12	.13	1.45	3.1	1.77	4.5
Responds to Cues	-1.30	.16	.80	-1.4	.76	-1.5
Engaged	-1.32	.14	.93	-.6	.92	-.6

MSq = mean square; STD = standardized residual.
[a]Did not conform to the expectations of Rasch Analysis.

Of the 3,749 responses to individual items, only 159 (4.2%) failed to conform to the expectations of the Rasch model. Thus data from 95.8% of responses "fit the model." This figure is well within the acceptable range.

The order of difficulty of TELS items is shown in Table 13.1. The items appeared to form a logical sequence. The items that represent "freedom from some constraints of reality" were the hardest and items that represent "framing" appeared to be the easiest, with the exception of "giving cues." Items that represent "internal control" and "intrinsic motivation" were more scattered. We will discuss the logic of this sequence below.

Respondents were asked to rate their own playfulness on a scale from "0" (not very playful) to "3" (very playful). No respondents assigned themselves a score of "0." Respondents' TELS scores ranged from -.48 to 2.96 logits with a mean of .92. Respondents who rated themselves a "1" ($n = 16$) had a TELS mean score of .38. Respondents who rated themselves a "2" ($n = 89$) had a TELS mean score of .83. Respondents who rated themselves a "3" ($n = 75$) had a TELS mean score of 1.16. When TELS scores were subjected to a single-factor ANOVA and Tukey Post Hoc tests (df = 2), there were significant differences among groups defined by respondents' self ratings [$F = 17.543, p = 0.00$]. Group 1's mean was significantly lower than either of the other groups and Group 2's mean was significantly lower than Group 3.

To address reliability of the instrument, we examined fit of respondents' data to the Rasch model and the reliability with which the assessment separated respondents into levels of leisure involvement.

The person separation index (Gp) was 1.47 indicating that this tool separated respondents into 2.29 levels of leisure involvement (4[Gp+1]/3). The reliability coefficient associated with the separation value (equivalent to Cronbach's alpha) was .68. The error for each item ranged between .09 and .16 with a mean of .11. No error exceeded the greatest desired value of .25.

DISCUSSION

Since leisure has been shown to improve quality of life and productivity (Cunningham, 1989; Shaffer, Duszynski & Thomas, 1982), an assessment that describes leisure experience is important to helping professionals. The TELS shows good potential to become a usable tool.

We examined goodness of fit of responses, items, and respondents; logic of the hierarchy sequence of items; item error estimates; levels of separation distinguished among respondents and items; and differences in scores among groups who rated themselves differently for playfulness (approach) as evidence of TELS (experience) validity and reliability. Several findings are of interest. First, there was a statistically significant difference in TELS scores among respondents who rated their playfulness a "1," "2,"or "3." Additionally, the items separated into approximately 11 levels, suggesting that the TELS has the potential to distinguish improvements that occur as a result of therapeutic intervention. Item error estimates also were very low suggesting item reliability.

Further, TELS items appeared to form a logical hierarchy representing one's experience within a serious leisure activity. The four hardest items ("pretends" extent and skill, "humor," and "imaginative use of objects") represent "freedom from some constraints of reality." These items also are among the most difficult for children on the ToP (See Appendix.B). These items may be harder because of unspoken standards in society for adult behavior. For example, most people respond well to playful mischief or teasing, which was much easier for adults than all other "freedom from some constraints of reality" items, than to someone pretending to be someone else. Often, we are cautious around someone we suspect is not being entirely honest with us. Thus, when adults pretend, they are likely to do so alone or in situations they consider to be very safe. Interestingly, although many of the items are not typically considered "adult behavior," they fit within the construct defined by the TELS, suggesting that these respondents recognized and engaged in them.

The two easiest items on the TELS represented "motivation" and "framing." (See Appendix. B.) Because respondents were asked to complete the TELS in the context of an activity in which they become intensely involved, it is logical that "engaged" was the easiest item. Although "Responds to cues" is moderately hard on the ToP, it was very easy on the TELS and it seems logical that responding to cues is much easier for adults than for children. With age, comes experience reading others' cues. In fact, adults have mastered cues so well that we are often manipulated by others feigning emotion (Csikszentmihalyi, 1997). In contrast, "giving cues" is much harder. Perhaps we are much more aware that we respond to others' cues but are less aware that we are also giving non-verbal cues to others.

The items in the middle of the scale are primarily items from the construct of "internal control" and "intrinsic motivation." Because we also use most of these skills ("social engagement," "social role," "shares," "supports," "enters," "negotiates") in many contexts and thus practice them frequently), it seems logical that they are moderately easy. For example, negotiating is a skill used to make a sale or to set up appointments. Many of these items have similar levels of difficulty, which leads to the question of whether all are necessary to the TELS. However, with the possible exception of "simultaneous activity," discussed below, each seems to provide important information not gained in any other question.

While the TELS shows promise, our results suggest the need for some changes and for further research. Specifically, changes need to be made to at least two items and the validity and reliability of the revised TELS should be re-examined with a more heterogeneous sample.

Perhaps the most significant difficulty we uncovered with the TELS was that data from 21 respondents (12%) did not conform to the expectations of the Rasch model. Of these, 11 had unacceptably low fit statistics, suggesting that their responses lacked variability. When these respondents' answers were examined, we found that 2 of them left 11 questions blank and the remaining 9 assigned the same response to approximately half of the questions. Lack of variability in respondents' data reflected a violation of the Rasch assumption that people more intensely involved in their leisure are more apt to get higher score on harder items than respondents who were less involved. These findings may suggest the need to create more difficult items. However, given that the TELS is designed primarily for use with people seeking the services of helping professionals, our finding may simply suggest the need for testing with a different population.

Of the 10 respondents with high fit statistics, 5 chose atypical activities such as work or reading schoolwork. High fit statistics indicated that answers were erratic; that is respondents scored themselves low on some easy items or high on some hard items. When the respondents' answers were examined in the context of their chosen activity, the reason for these erratic responses seemed clearer. For example, in the context of being at work, it is reasonable that one does the work as much for pay or recognition as for the process. Also, it probably is not advisable to modify work assignments. Likewise, when coaching children in gymnastics (one respondent's choice), the coach often tries not to choose the direction of activities; rather, a coach may need to be more supportive of participants' ideas, giving guidance instead of direction. Though Csikszentmihalyi (1997; Csikszentmihalyi & Kleiber, 1991) indicated that intense involvement could occur in many types of activities, not simply leisure, it may be that this instrument is not appropriate to apply to experiences linked to all activities in which an individual experiences intense involvement. Therefore, future researchers using the TELS should consider limiting respondents' choice of context to leisure activities. However, it is important to note that data from several respondents who chose atypical leisure activities such as work did conform to the expectations of the Rasch model.

Another difficulty we encountered was that, while the items separated into 11 levels, they only distinguished slightly more than 2 levels of leisure experience in the respondents. Lack of separation potentially reveals a problem of reliability. However, it seems likely that the small separation index is the result of a relatively homogeneous sample, of which 95% were college students.

As was expected, the respondents' TELS scores formed a bell curve. Yet the comparison of the respondents' curve to the bell curve set by the items shows another difficulty with these results. The mean for the respondents was approximately +1.0 logits. This indicates either the TELS is an easy tool, that the sample is highly involved in their leisure, or both. Most (75%) of the responses were "2s" or "3s." The apparent involvement in leisure of the sample is not a surprise, given that 95% were college students. However, because the TELS was designed to assist with assessment and intervention planning, these results demonstrate the need for future research with a more heterogeneous population.

Another difficulty we encountered with the TELS was that two skill items, "simultaneous activity" and "social role," failed to fit the Rasch model. "Simultaneous activity" asks how many other activities respondents typically were involved in at the same time they were doing the primary activity. However, in retrospect, the answer to this question seemed somewhat dependent on the type of primary activity selected. For example, when playing an organized sport (e.g., volleyball), it is unlikely that a respondent would be involved in any other activity. In contrast, activities such as watching television and socializing often "pull" for simultaneous engagement in other activities. Adults multi-task to increase their arousal state, making a mundane activity interesting (Blanche, 1999). Further, they may choose repetitive activities to free their minds for other activity. Therefore, being engaged in one activity at a time is not necessarily easier for more people who are more involved in their leisure and harder for people who are less involved. Thus, "simultaneous activity" appears to violate a basic assumption of the Rasch model: easy items are easy for all people. In future research, this item probably should be removed.

"Social role" asks whether respondents see themselves as a leader, an equal participant, slightly less an equal and more a follower, or definitely a follower while engaged in their chosen activity. Almost all respondents indicated that they were the leader ("3") or an equal participant ("2"). This small variability in scoring explains why "social role" had low fit statistics. With this homogeneous sample, data from "social role" did not conform to the assumptions of the Rasch model. Therefore, this item should be monitored closely in future research to determine whether or not it remains as an item of the TELS.

SUGGESTIONS FOR FUTURE RESEARCH

While the TELS shows promise for use by helping professionals, several changes seem important. First, the items "simultaneous activity" and "social role" should be modified significantly or removed from the TELS, as they do not appear to be a part of the construct of leisure experience defined by the TELS.

We recommend that a revised TELS be used with a sample diverse in age and anticipated involvement in leisure and comprised, at least in part, of people likely to seek the services of helping professionals. An important question to be addressed by such research is whether the tool is able to distinguish more levels of leisure involvement among the respondents. We also recommend that future researchers examine the relationship of demographic information to TELS scores. The addition of demographic questions such as socioeconomic status might shed additional light on the leisure experience of adults since Griffith (2000) found a correlation between mother's socioeconomic status and playfulness in children.

REFERENCES

Blanche, E. I. (1998). *Play and process: The experience of play in the life of the adult.* Unpublished doctoral dissertation, University of Southern California, Los Angeles, CA.

Blatner, A. & Blatner, A. (1988). *The art of play, an adult's guide to reclaiming imagination and spontaneity.* New York: Human Sciences.

Brightbill, C. K. (1960). *The challenge of leisure.* Englewood Cliffs, NJ: Prentice – Hall.

Bundy, A. C., Nelson, L., Metzger, M. & Bingaman, K. (2001). Validity and reliability of the test of playfulness. *Occupational Therapy Journal of Research, 21,* 276-292.

Bundy, A. C. (2000). *Test of playfulness (Revised version 3.5) manual.* Ft. Collins, CO: Colorado State University.

Bundy, A. C. (1993). Assessment of play and leisure: Delineation of the problem. *The American Journal of Occupational Therapy, 47*(3), 217-222.

Cohen, D. (1993). *The development of play.* New York: Routledge.

Csikszentmihalyi, M. (1997). *Finding Flow.* New York: Basic.

Csikszentmihalyi, M. & Kleiber, D. A. (1991). Leisure and self-actualization. In Driver, B.L., Brown, P.J., & Peterson, G.L., (Eds.). *Benefits of Leisure* (pp. 91-102). State College, PA: Venture.

Cunningham, P. H. (1989). Stress in relation to satisfaction with leisure experienced by those performing in therapeutic recreation. *Psychological Reports, 64,* 652.

Des Camp, K. D. & Thomas, C. C. (1993). Buffering nursing stress through play at work. *Western Journal of Nursing Research, 15,* 619-627.

Gaik, S. & Rigby, P. (1994). *A pilot study to address the reliability and validity of the test of playfulness—research version 2.2 and to compare the playfulness of children with physical disabilities with age-matched able-bodied peers. Research Report.* Hamilton, Ontario: McMaster University, Neurodevelopmental Clinical Research Unit.

Gibbs, N. (1989). How America has run out of time. *Living, April 24,* 58-67.

Gliner, J. A. & Morgan, G. A. (1998). *Research design and analysis in applied settings.* Fort Collins, CO: Colorado State University, (pp. 139-163).

Glynn, M. A. (1994). Effects of work task cues and play task cues on information processing, judgment, and motivation. *Journal of Applied Psychology, 79,* 34-45.

Glynn, M. A. & Webster, J. (1992). The adult playfulness scale: An initial assessment. *Psychological Reports, 71,* 83-103.

Griffith, L. R. (2000). *Hispanic American children and the test of playfulness.* Unpublished master's thesis, Colorado State University, Fort Collins, CO.

Harkness, L. & Bundy, A. C. (2001). The test of playfulness and children with physical disabilities. *Occupational Therapy Journal of Research, 21,* 73-89.

Hensley, W. E. & Roberts, M. K. (1976). Dimensions of Rosenberg's scale of self esteem. *Psychological Reports, 38,* 347-357.

Hogg, J. (1993). Creative, personal and social engagement in the later years: Realisation through leisure. *The Irish Journal of Psychology, 14,* 204-218.

Kerr, J. H. & Apter, M. J. (1991). *Adult play.* Amsterdam/Lisse: Swets & Seitlinger B.V.

Leipold, E. & Bundy, A. C. (2000). Playfulness and children with ADHD. *Occupational Therapy Journal of Research, 20,* 61-82.

Linacre, J. M. & Wright, B. D. (1991-1994). *A user's guide to BIGSTEPS, Rasch-model computer program.* Chicago, IL: MESA Press.

Mannell, R. C. (1993). High-investment activity and life satisfaction among older adults: Committed, serious leisure, and flow activities. In J. R. Kelly, (Ed.), *Activity and Aging* (pp.125-145). Newbury Park, CA: Sage.

Matsutsuyu, J. S. (1969). Interest checklist. *American Journal of Occupational Therapy, 23,* 323-328.

Murgatroyd, S., Rushton, C., Apter, M. & Ray, C. (1978). The development of the Telic Dominance Scale. *Journal of Personality Assessment, 42,* 519-528.

Neulinger, J. (1974). *The psychology of leisure: Research approaches to the study of leisure.* Springfield, IL: Charles. C. Thomas.

Neumann, E. A. (1971). *The elements of play.* New York: MSS Information.

Porter, C. A. & Bundy, A. C. (2000). Validity and reliability of three tests of playfulness with African American children and their parents. In S. Reifel (Ed.), *Play and Culture Studies* (Vol. 3): *Theory In Context and Out* (pp. 315-334). Westport, CT: Ablex.

Previd, D. (1999). *Validity and reliability of the test of playfulness version three.* Unpublished master's thesis. Colorado State University, Fort Collins, CO.

Primeau, L. A. (1995). Work and leisure: Transcending the dichotomy. *American Journal of Occupational Therapy, 50,* 569-577.

Schaefer, C. (Ed.). (1993). *The therapeutic powers of play.* Northvale, NJ: Jason Aronson.

Shaffer, J. W., Duszynski, K. R. & Thomas, C. B. (1982). Youthful habits of work and recreation and later cancer among physicians. *Journal of Clinical Psychology, 38,* 893-900.

Sutton-Smith, B. (1997). *The ambiguity of play.* Cambridge, MA: Harvard.

Sutton-Smith, B. (1977). Play as adaptive potentiation. In P. Stevens, (Ed.), *Studies in the anthropology of play.* Cornwall, NY: Leisure.

Tyler, R. (1996). *Girls, boys, and a top: Gender/environmental differences and a Test of Playfulness.* Unpublished master's thesis, Colorado State University, Fort Collins.

Witt, P. A. & Ellis, G. D. (1989). *The leisure diagnostic battery user's manual.* State College, PA: Venture.

APPENDIX A
TELS Protocol
(Demographic questions not included)

Playfulness is a way of approaching tasks flexibly, with the knowledge that there is no correct way to do them. We are interested in your perceptions of your playfulness, when you are doing an activity in which you become so involved that you are not aware of other things going on around you. While we think many activities in which you become totally engaged will be leisure or play activities, some may be work, self care, or other kinds of tasks. Please take about 10 minutes to fill out this survey. Consider your answers carefully.

1) While filling out this survey think of the activity in which you become most intensely involved.

Name that activity here:

During an activity in which you become totally involved, which of the following best describes your actions? If the activity you are thinking of is done alone, you can skip questions 14-24.

I would . . .

Be intensely involved for

more than 1 hour
1/2 hour – 1 hour
15 min. - 1/2 hour
less than 15 min.

Be involved in

only one activity at a time
two activities at the same time
more than 2 activities at the same time

Decide what I would do and how I would act rather than letting someone else decide.

almost all the time
much of the time
some of the time
rarely or never

Feel physically and emotionally safe.

almost all the time
much of the time
some of the time
rarely or never

Continued on next page

APPENDIX A - *Continued*

Demonstrate real exuberance (giggling, laughing out loud, singing,
shouting)

almost all the time
much of the time
some of the time
rarely or never

Try to overcome anything that makes
the activity difficult rather than
switching to something else

very intensely
moderately intensely
mildly intensely
not intensely

Modify the activity a little to maintain
the fun or the challenge. I would
change it:

spontaneously and effortlessly
once in a while or with some difficulty
with great difficulty or I would go
along when someone else changed it.
not at all.

Engage in playful teasing or mischief
in order to engage another or have
more fun.

almost all the time
much of the time
some of the time
rarely or never

Characterize the playful teasing or
mischief that I do as
particularly clever.
a little clumsy or stilted.
sometimes a little offensive to others.
often offensive to others.

Engage in the activity for the sheer
pleasure of it rather than because I
want to win or gain approval or some
other tangible reward.

almost all the time
much of the time
some of the time
rarely or never

Continued on next page

APPENDIX A – *Continued*

Pretend to be: someone else / to be   Pretend in a way others would find
doing something I am not doing / or
that an object is something it is not.

almost all the time   highly convincing
much of the time   moderately convincing
some of the time   mildly convincing
rarely or never   not convincing at all
                                           not applicable (if you selected d on
                                           question 11)

Incorporate people, objects, or activity   Negotiate to get my needs met
in imaginative ways.

almost all the time   with skilled give and take
much of the time   in a moderately skilled way
some of the time   by being bossy
rarely or never   would not try

Involve others in the activity   Characterize my role with others as

almost all the time   the leader
much of the time   an equal participant
some of the time   slightly more a follower than an equal
rarely or never   definitely a follower
                                           not applicable (if you selected d on
                                           question 15)

Do things that help another enjoy the   Enter a group already involved in an
activity   activity

almost all the time   by blending in effortlessly
much of the time   a bit awkwardly
some of the time   by waiting for an invitation
rarely or never   in a very unskilled way

Continued on next page

APPENDIX A – *Continued*

Initiate activity with another person

effortlessly
a bit awkwardly
by waiting for an invitation
in a very unskilled way

Tell jokes or funny stories or engage
in exaggerated behavior to get

attention
almost all the time
much of the time
some of the time
rarely or never

Share belongings, friends, ideas, space

almost all the time
much of the time
some of the time
rarely or never

Give clear cues about how others
should interact with me

almost all the time
much of the time
some of the time
rarely or never

Respond to others' cues in a way that
furthers the activity

easily and effortlessly
a little awkwardly
somewhat awkwardly
very awkwardly

Continued on next page

APPENDIX A – *Continued*

Be intensely involved for more than 1 hour 1/2 hour – 1 hour 15 min. - 1/2 hour less than 15 min.	Be involved in only one activity at a time two activities at the same time more than 2 activities at the same time

---

Decide what I would do and how I would act rather than letting someone else decide. almost all the time much of the time some of the time rarely or never	Feel physically and emotionally safe. almost all the time much of the time some of the time rarely or never

---

Demonstrate real exuberance (giggling, laughing out loud, singing, shouting) almost all the time much of the time some of the time rarely or never	Try to overcome anything that makes the activity difficult rather than switching to something else very intensely moderately intensely mildly intensely not intensely

---

Modify the activity a little to maintain the fun or the challenge. I would change it: spontaneously and effortlessly once in a while or with some difficulty with great difficulty or I would go along when someone else changed it. not at all.	Engage in playful teasing or mischief in order to engage another or have more fun. almost all the time much of the time some of the time rarely or never

---

Continued on next page

APPENDIX A – *Continued*

Characterize the playful teasing or mischief that I do as
particularly clever.
a little clumsy or stilted.
sometimes a little offensive to others.
often offensive to others.

Engage in the activity for the sheer pleasure of it rather than because I want to win or gain approval or some other tangible reward.
almost all the time
much of the time
some of the time
rarely or never

Pretend to be: someone else / to be doing something I am not doing / or that an object is something it is not.
almost all the time
much of the time
some of the time
rarely or never

Pretend in a way others would find
highly convincing
moderately convincing
mildly convincing
not convincing at all
not applicable (if you selected d on question 11)

Incorporate people, objects, or activity in imaginative ways.
almost all the time
much of the time
some of the time
rarely or never

Negotiate to get my needs met
with skilled give and take
in a moderately skilled way
by being bossy
would not try

Involve others in the activity
almost all the time
much of the time
some of the time
rarely or never

Characterize my role with others as
the leader
an equal participant
slightly more a follower than an equal
definitely a follower
not applicable (if you selected d on question 15)

Continued on next page

APPENDIX A – *Continued*

Do things that help another enjoy the activity
almost all the time
much of the time
some of the time
rarely or never

Enter a group already involved in an activity
by blending in effortlessly
a bit awkwardly
by waiting for an invitation
in a very unskilled way

Initiate activity with another person
effortlessly
a bit awkwardly
by waiting for an invitation
in a very unskilled way

Tell jokes or funny stories or engage in exaggerated behavior to get attention
almost all the time
much of the time
some of the time
rarely or never

Share belongings, friends, ideas, space
almost all the time
much of the time
some of the time
rarely or never

Give clear cues about how others should interact with me
almost all the time
much of the time
some of the time
rarely or never

Respond to others' cues in a way that furthers the activity
easily and effortlessly
a little awkwardly
somewhat awkwardly
very awkwardly

How playful do you consider yourself to be? (circle one number)

Not Playful at all   Very Playful

0   1   2   3

APPENDIX B
Comparison of TELS and ToP Item Order

TELS   **ToP**

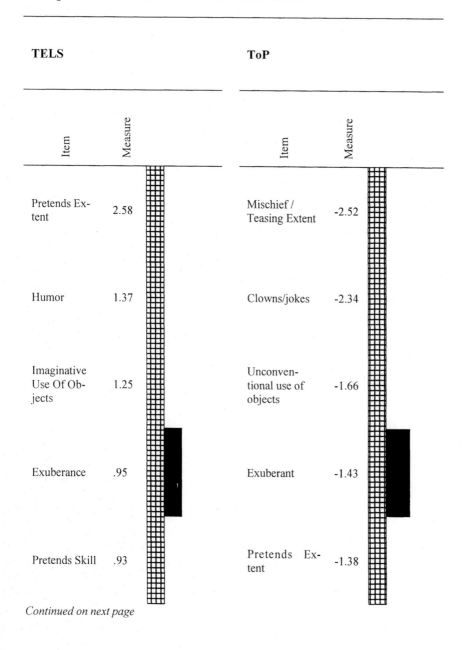

Item	Measure		Item	Measure
Pretends Extent	2.58		Mischief / Teasing Extent	-2.52
Humor	1.37		Clowns/jokes	-2.34
Imaginative Use Of Objects	1.25		Unconventional use of objects	-1.66
Exuberance	.95		Exuberant	-1.43
Pretends Skill	.93		Pretends Extent	-1.38

Continued on next page

APPENDIX B – *Continued*

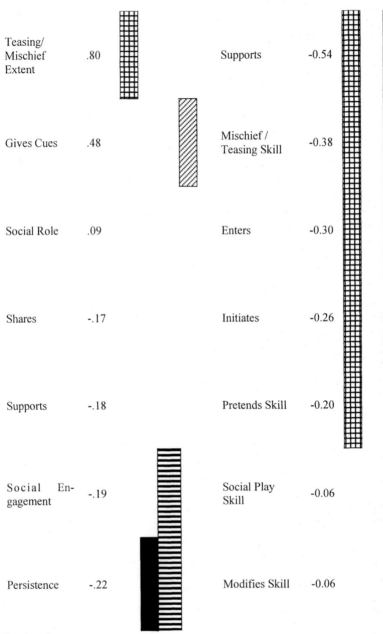

Teasing/ Mischief Extent	.80	Supports	-0.54
Gives Cues	.48	Mischief / Teasing Skill	-0.38
Social Role	.09	Enters	-0.30
Shares	-.17	Initiates	-0.26
Supports	-.18	Pretends Skill	-0.20
Social Engagement	-.19	Social Play Skill	-0.06
Persistence	-.22	Modifies Skill	-0.06

Continued on next page

APPENDIX B – *Continued*

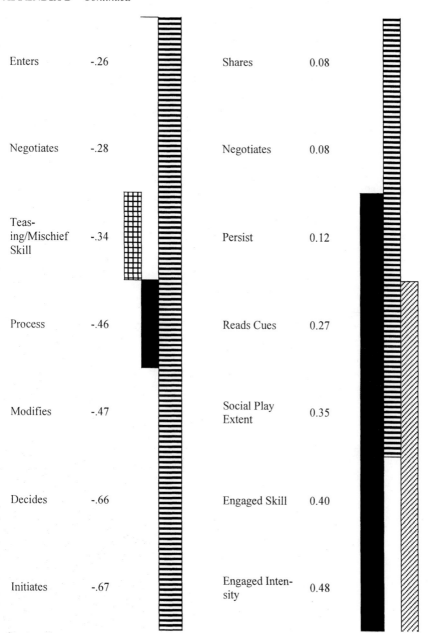

Enters	-.26
Negotiates	-.28
Teas-ing/Mischief Skill	-.34
Process	-.46
Modifies	-.47
Decides	-.66
Initiates	-.67

Shares	0.08
Negotiates	0.08
Persist	0.12
Reads Cues	0.27
Social Play Extent	0.35
Engaged Skill	0.40
Engaged Inten-sity	0.48

Continued on next page
APPENDIX B – *Continued*

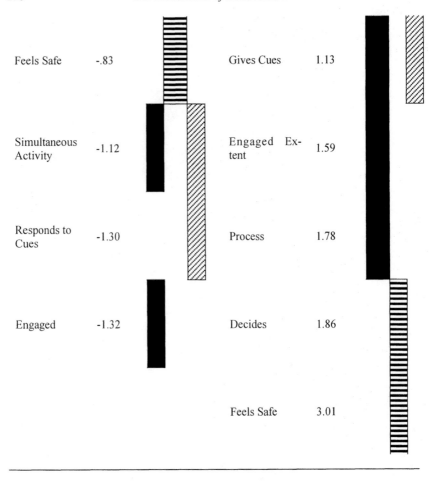

Feels Safe	-.83		Gives Cues	1.13
Simultaneous Activity	-1.12		Engaged Extent	1.59
Responds to Cues	-1.30		Process	1.78
Engaged	-1.32		Decides	1.86
			Feels Safe	3.01

Note: ⊞ = Reality   ■ = Motivation   ≡ = Control   ▨ = Framing

The Facets program used for the TELS states that hardest to easiest items descend from the highest positive score to the lowest negative score. Conversely, the BIGSTEPS program used for the ToP states that hardest to easiest items descend from the lowest negative score to the highest positive score.

Chapter 14

Collecting as a Form of Play

Paola de Sanctis Ricciardone

The purpose of this article is not to claim that all kinds of collecting are forms of play but that sometimes—above all in individual behaviors—some features of gathering objects and constructing special and protected worlds can be perceived as a ludic activity. After a brief introduction about some problems linked to the uses of the terms play and game, this will be an attempt to confront the relationship between the study of play and some fruitful contaminations with the philosophy of language. Then, the distinction between *mémoire volontaire* and *mémoire involontaire* elaborated by Walter Benjamin (1982). (to some extent connected to the previous questions) will be used as a metaphor for representing different and sometimes conflicting strategies in collecting. In the middle of the XVIII century the shift from the individualistic, purposeless and playful practice of collecting to a centralist and institutional one, was not without its birth pains. Some "old fashioned" collectors were victims of mockery and irony in all Europe. In the final paragraph will be presented some examples of contemporary collecting practices, drawn also from personal research, to demonstrate that despite condemnations and ridicule the sense of wonder and playfulness in collecting is still alive and well.

Issues of translation + dfn:

HAPPY AND UNLUCKY PEOPLE

play vs game → English distinguishes

To say, "play," is to be aware that some clarification is needed. There are several definitions of "play" in the specialist literature to which TASP has contributed in influential ways, most recently; Sutton-Smith (1999) and Henricks (1999). The different meanings between play and game in native English language were and still are very useful as a distinction to increase the theoretical underpinnings in different fields, from G. H. Mead (1896) to von Neumann and Morgenstern (1944) or Avedon and Sutton-Smith (1971).

One can imagine the difficulties that English translators encountered when they faced the two holy texts for the study of play: Huizinga's *Homo Ludens* and Caillois *Le Jeux et les Hommes*, and even in translating the notion of *Spraches-piel* of Wittgenstein. Additionally, it is possible to read in Allen Sack (1977, p. 186), John Bowman (1978, p. 241), or Janet Harris (1981) about their uneasiness with the way Huizinga and Caillois used (or misused) the terms, play and game. But the problem is only inherent in the translation. Italian, French, Dutch or German languages, have only one term to express both play and game, in turn *'Gioco,' 'Jeu,' 'Spel,' 'Spiel.'* That fact makes these scholars very happy when talking about play and game in everyday life: They don't have problems of choice or risk mistake. They are, instead, less happy, and to some extent, un-lucky when having to shift to other levels of reflections, in observing, studying, defining, interpreting, sometimes explaining this nebulous and strange, odd, or quirky—as Sutton-Smith says (1999, p. 245)—human and non-human activity.

Caillois is less French-centric than one would think. His invention of the two tensions that drive the players—namely *paidia* and *ludus,* "ways of playing" as he said—is a good tool to render the difference, omitted in the French term *Jeu,* between a turbulent, disorderly playful activity and a ruled, organized and socially rooted one.

For von Neumann and Morgerstern (German-centric?) the terms play and game are two of the *Theory of Game's* "Termini Technici," forming a sort of hierarchy. These terms don't indicate two different styles of playing but two dif-ferent levels of conceptualizing the field of study, one more abstract and the other more concrete. "The game—as it is read—is simply the totality of the rules which describe it. Every particular instance at which the game is played—in a particular way—from the beginning to end, is a play" (von Neumann & Morgenstern, 1944, p. 49). Therefore, the game is an "abstract concept" and play is one of the endless concrete, indexical reifications generated by the rules of such a game. This difference recalls the difference between *langue* and *parole* elaborated by Ferdinand de Saussure in his *Course in General Linguistics* in which he often uses the chess game as a perfect metaphor of language. Accord-ing to de Saussure (1973), *langue* is the whole set of super-individual rules (a "social fact") and *parole* is the concrete individual act of speaking (the chess move) generated by the rules the speaker (player) knows.

That of von Neumann and Morgenstern is a strong definition, ascribing to one philosophical perspective that rejects conceptual indeterminacy and skepti-cism about knowledge (Sassower, 1995). As a matter of fact it seems that play without a generating (and intelligible) ruled game is not possible in the *Theory of Games and Economic Behavior.*

The same kind of problems will be faced also in parallel and more recent play literature, as summarized by Linda Hughes (1983). In her article the ques-tion about the rules of an abstract game and the actual event of playing is an-swered by adopting the term gaming, "to distinguish between the game itself and what the real players do in constructing a particular instance of that game" (Hughes, 1983, p. 189).

CHESS GAME, BALL PLAY AND LANGUAGE

Almost in parallel to von Neumann and Morgenstern, Wittgenstein in his *Philosophical Investigations* (1953), started in 1941 and concluded in 1949, seems to handle at the highest level the long fruitful contamination between play and language. But in a very different way from the *Theory of Games*, as Lyotard suggests (1979, p. 23). Both the notions of *Spiel* and the consequent *Sprachspiel* in Wittgenstein can't be "mastered" by one definition or rules. It is not by chance that Wittgenstein opted to neglect his preferred chess metaphor in favor of a more fluid metaphor found in paragraph 83. He needed to show the enlightening analogy "between *Spiel* and language." So he uses an imaginary, strange, quirky ball: *Spiel* (better play in this case) between men who enjoy throwing the ball to each other in a very odd and changeable manner without given goals, predictable outcomes and settled rules. Wittgenstein seems to suggest to us that it isn't worth trying to find different rules in every single throw (elsewhere, in paragraphs 81 and 82, he talks about the possibility, in the *Sprachespiele*, that rules can be missed or hidden to the same speaker-player who acts a discursive sequence). Therefore, it is better to try to grasp the whole flow of such strange playing, with its internal blurred (and often not drawn at all) borders. Maybe that in playing, we "make up the rules as we go along" and change the rules "as we go along," as he says in English (p. 83). That is a weak definition of *Spiel*; we either want to refer to play or game. Instead, in the *Theory of Games and Economic Behavior*: "The rules of the game are absolute commands. If they are ever infringed, then the whole transaction by definition ceases to be the game described by those rules" (von Neumann & Morgenstern, 1944, p. 49).

To the contrary, in the case shown by Wittgenstein, the young men change and break, perhaps ignore rules, but the ideal transaction between them doesn't cease for the simple reason that they keep on playing. The Spiel they are playing is not described by agreed-upon rules. The only definition we can give in this case is after the fact, to evoke Clifford Geertz (1995); that is to say we can find rules after the game is done (Wittgenstein, 1953). Such a form of feedback loop, searching everywhere a rationale, was one of the favorite procedures of past anthropology. Today, more often, anthropologists (see i.e. Clifford & Marcus, 1986; Borofsky, 1994) prefer to surrender to the evidence and accept that sometimes—though not always—human behaviors can lack driving rules or underlying rationale.

All such a boring digression to say that we can find even in the history and studies of collecting, many of the problems raised in studies of play.

MEMORIES IN CONFLICT

Walter Benjamin in his *Das Passagen Werk* (1982, pp. 269-280) addressed his attention to Huizinga's concept of Magic Circle.

The true collector for Benjamin inserts objects into a safe and protected space, freeing them from all utilitarian connection. The new type of connections between objects invented by the collector lies on a special kind of memory or reverie, defined in French terms by Benjamin's *mémoire volontaire* (involuntary, unintentional memory). The 'canon' of this *mémoire volontaire* is 'a certain productive disorder.' The *mémoire volontaire* is instead a sort of register that provides the object with an order's number behind which [the object] vanishes (1982, p. 280).

And this recalls something that can happen to the objects in the more traditional museums with their aseptic labeling systems.

The *mémoire involontaire*, as conceived by Benjamin seems to produce the same "productive disorder" of the strange, odd and quirky ball play depicted in the *Philosophical Investigations* (Wittgenstein, 1953). The rules of the collector/player, if they are any, can be hidden, eluded, changed "as we go along" (Wittgenstein, 1953, p. 83). The *mémoire volontaire*, instead, with its more rigid strategy descending from a previous, intentional project, seems to produce something similar to a game.

Games are repeatable because of their systematic pattern and their predictable outcomes. Play on the other hand is less systematic, and is open-ended with respect to outcomes. In a game, the participant's voluntary control over procedures has been subordinated in anticipation of, but without guarantee for, a given goal (Avedon & Sutton-Smith, 1971).

Moreover, in the constructs produced by the *mémoire volontaire* it is possible to find the whole range of this type of continuum whose ends are the play and the playful (Sutton-Smith, 1997), "where play [is] contained by frames and playful [is] disruptive of frames" (p. 196).

PRODUCTIVE DISORDER AND *ORDRE MÉTHODIQUE*

In his writings about the collector, and the difference between *mémoire volontaire* and *volontaire*, Benjamin seems to grasp and summarize the core of a long controversy. In the western history of collecting it seems the shift is important—in the middle of 18th century—from the individualistic and diffused practices to the institutional and centralist ones, with the birth of the more important European national museums, like the British Museum for example (MacGregor, 1985).

Emblematic of this shift is the strong and ironic "death-sentence" that Diderot and D'Alembert (in the second tome of the *Encyclopédie*, 1757) passed against every form of naive and purposeless collecting. There are three entries particularly interesting, namely *Bibliomane*, *Bibliomanie*, and *Cabinet d'Histoire Naturelle* (Bibliomania, Bibliomaniac and Cabinet of Natural History). Here the French Encyclopedists seem to serve an obituary of the eclectic, dreaming and wondering collector, with his obsolete *Kunst- und Wunderkammern*, Art and Wonder Cabinets. They state the primacy of the spiritual pleasures on the visual pleasures: "*L'ordre méthodique [. . .] n'est presque jamais celui qui est le plus advantageux aux yeux*" (the methodic order is never the most suitable to eyes, s.v. *Cabinet d'histoire naturelle*).

Actually many of these kinds of collectors were already important in previous centuries. The great "collecting machine" (Lugli, 1983, p. 94) started in all of Europe after the 14th century. This machine assembled, preserved and saved a huge amount of artistic, bibliographical, archeological, naturalistic goods, rarities and curiosities. In many cases that disseminated private patrimony stands at the beginnings of the most prestigious public national museums (Pearce, 1994). Good evidence for this process was the life of the legendary British collector Hans Sloane (1660-1753), probably the last heir of this old "collecting machine" (Alexander, 1996; MacGregor, 1985). MacGregor says that it "takes us beyond the end of the seventeenth century and beyond the point where the term 'Cabinet of curiosities' has any useful meaning" (p. 157-158).

When MacGregor died, he left, for only 20.000 pounds, his huge eclectic collections to the British Crown. So in 1759, thanks also to a special lottery arranged by the Parliament raising money to sustain the offer, the British Museum was born.

But Diderot and D'Alembert raised the problem of the rules of collecting, the rational (and "ordered") way of gathering, selecting and displaying objects. For the *Philosophes* the driving principles for a good, educational and useful Museum must be external to a single individual, neutral and in accordance to the criteria of the institutional scientific and historical knowledge. In other terms it seems that from the middle of the 18th century a sort of argument started between *mémoire volontaire* and *mémoire involontaire* that can be reconnected even to the different meanings we can give to the terms game and play.

The public and institutional counterpart of collecting and displaying objects—in the aims of theorists—must be "mastered" by definite games, and only the specialists in every field of knowledge can settle and legitimize the rules of these games. Also for the ethnological collections, the "Museum Era" (from 1840s to 1890) was established quite in accordance to the main paradigms of the anthropological theories (Stocking, 1985).

ENGLISH AND VENETIAN MOCKERIES

Even in the popular and public opinion we can find clues of the same suspicious and ironic attitude about the private collectors. In his long life, Hans Sloane was esteemed and honored by sovereigns, scientists, and famous intellectuals. But he couldn't escape from the fun that many contemporaries had with him. Poets and humorists wrote satirical verse about his "knick knackatory" and his frenzy for strange and rare objects (Alexander, 1996). For example, James Salter authored a true masterpiece of mockery, not too far from the austere Chelsea's home-museum of Sir Sloane:

> [He] sets up a comic museum in his Don Saltero's coffeehouse in Chelsea. It contained an asbestos purse from North America that Franklin had sold Sloane but also such fakery as 'the Queen of Sheba's Fan and Cordial Bottle' and 'Pontius Pilate's Wife's Chambermaid's Sister's Hat. Saltero printed a satirical catalogue of his collection (Alexander, 1996, p. 45).

In the same period, even in Italy, evidence is found of a similar satirical attitude toward the collectors.

The Carlo Goldoni comedy of manners *La Famiglia dell'Antiquario* (*The Family of the Antiquary*) was performed for the first time in Venice in 1750. Here the target of the severe irony of Goldoni is a Venetian Count, Anselmo Terrazzani. Anselmo is completely devoted to cultivating his passion for antiques, archeological relics and rarities, an exclusive fashion in the period of the Grand Tours. But he doesn't have any specific knowledge about this stuff. His desire to possess rare and unique objects is so strong that he becomes victim to all Venetian cheats. Harlequin and Brighella, the two famous Italian servant-masks, have the task (a quite imperative one) of taking advantage of the insane and blind frenzy of their lord, palming off every sort of knick-knack to him. The "collection" of Count Anselmo is actually an incredible gallery of dearly paid fakeries.

Pantaloon, the classic Venetian merchant-mask, is assigned the task embodying bourgeois common sense, solid, frugal and far from all aristocratic extravaganzas. In this duetto (Scene XVIII, Act first) Anselmo shows to Pantaloon the last ingenious swindle acted by Harlequin and Brighella:

ANSELMO: Good morning, my dear friend. You that are a merchant, man of the world and connoisseur of rarities, please value for me this beautiful antiquity.

PANTALOON: What a great opinion you have of me as a merchant, that you request I value an oil lamp.

ANSELMO: Poor mister Pantaloon, you don't know anything. This is the Eternal Light of the Sepulchre of Ptolemy.

PANTALOON: [Laughs]

ANSELMO: Yes, of Ptolemy, found in one of the Egyptian pyramids.

PANTALOON: [Laughs]

ANSELMO: You laugh because you aren't a connoisseur.

PANTALOON: Very well, I'm ignorant, and you are such an expert. I don't want to pick a quarrel with you about that. But I tell you that everybody in the town wonders how a lord like you can lose his time and his money for this sort of bullshit.

ANSELMO: Gossip of the envious. The same people that condemn me in public applaud me in private.

PANTALOON: There isn't anybody that envies your gallery, which consists of a wealth of rags. [...].

ANSELMO: In this world everyone has some kind of leisure. Someone plays, someone goes to the taverns, and I amuse myself with the antiquities.

Well, Anselmo is certainly a "victim" of all kinds of "false identification" which in this Goldoni comedy seems a grotesque and extreme case of the cultural *allodoxie* depicted by Bourdieu (1984, chap. 6), namely the heterodoxy perceived as orthodoxy, connected to a strong desire of distinction. Anyway, Anselmo Terrazzani remains a dreamer who has chosen collecting objects as a form of leisure and a way to be consumed by wonder.

In the following centuries the literature will show a plenty of evidences about collectors, seen often as characters signed by passion, but also by a sort of social liminality and inner inscrutable extravaganza. The novel *Le cousin Pons* (1994), that Balzac wrote in 1847, is a true masterpiece in this sense. The intriguing charm of the collectors still seduces contemporary intellectuals and writers (Rosenberg 1970; Chatwin 1988; Sontag 1993). This literature offers unique insights into collecting and strategies of identity construction, one of the most interesting themes of the contemporary "collection studies." Akin (1996) writes:

> Measuring the relative popularity of a type of collection within a society can indicate cultural norms, and the way people relate to the rest of society is often defined in terms of social acceptability of their collecting passions. Collectors define themselves, and are defined by others, in part, by what they collect. Obituary notices, for example, often mention the deceased's collecting passions (p.109).

WONDER IS STILL ALIVE AND WELL

Etherodoxy apart, the conflict between the two different tensions in collecting strategies, one more idiosyncratic, playful, and disordered (as in the ball-Spiel of Wittgenstein) and the other more ruled and useful, appears even inside of the museum's multiple and overlapping images. According to Greenblatt (1991, p.54), the two distinct and alternative "models" for museums—namely resonance and wonder—are inescapable in the past and still now "for both the poetics and the politics of representation."

Although the condemnations and sometimes damnation of what Benjamin called *mémoire involontaire*, a huge deal of evidence (literary, historical, from collection studies, from anthropology, from material culture studies, from fieldwork etc.) seems to show that playfulness and wonder are—still now—the major driving force of several collectors.

For many authors, such as Susan Stewart (1984, pp. 32, 151), Danet and Katriel (1989), and Susan Pearce (1993, pp. 50-55), the private side of collecting is a form of play. Generally it is a free and voluntary activity conducted during leisure time, even though some collectors define themselves ironically as "addicts" like Peggy Guggenheim (1997). In those works, collecting is a form of play "with classification." According to Susan Stewart and James Clifford (1988) (and before them Baudrillard, 1968), classification and knowledge about objects represent the shift from "an excessive, sometimes even rapacious need to have . . . into ruled governed, meaningful desire. Thus, the self that must possess but cannot have it all learns to select, order, classify in hierarchies" (Clifford, 1988, pp. 218-19).

Even in childhood, says Clifford, sooner or later the little collector will learn to display in a proper way dolls, model cars or dinosaurs. That makes the difference, above all for the adults, between a sane collecting and "the dark side" of collecting, says Pearce (1993, p. 51), namely the fetishism, the compulsiveness, the misery or the coprophilia.

As in the study of play (Bateson, 1971; 1978; Goffman, 1974, pp. 40-41) the question of the frame is central in the study of collecting. Inside the "magic circle" of a collection something happens to the objects. Even the most mundane and trivial ones become different from what they were in their original context and enhance their value, at least subjectively. For Appadurai this is the power of "the esthetic of diversion" (1986, p. 28).

The objects become semiophores, in terms of Pomian (1990), that is to say "meaning carriers." They cease to be useful and mere "things" and become material words of special discourses and esthetic constructs.

But this strong divide between things and semiophores, between useful and collectible objects seems sometimes to collapse when the behavior and the way of life of certain collectors are considered. Even the object's interpreting frame sometimes tends to blur and not only to the eyes of an external observer. Evoking Bateson the implicit message, "This is a collection," can be neglected by the same collector.

In my research I encountered and interviewed many Italian collectors. Some of whom are wealthy and important. One collects, with passionate care and capacity in selecting authentic items, precious objects, antiquities, archeological relics etc. In this case, generally their homes exhibit only a discretional amount of the whole collection. The rest is kept elsewhere; it may be in an armored room or in bank vaults. To some extent, the objects of great value are self-framed; to handle them is difficult and dangerous. For them the usefulness is truly "banished forever," as Benjamin says.

But there are other drifts in the contemporary practices of collecting that I encountered in specialist literature (Mullen, 1994; Windsor, 1994; Belk, 1994; Pearce, 1998; Pertinaz, 1996; Bloom, 1997) and in my research. The collections of extreme ephemera and minimalist items, objects we can find in everyday-life, such as beer-mats, tourist souvenirs, cans or telephone cards seem to be, as Susan Stewart suggests (1984, 167), "anticollections." In some cases the collector draws material or virtual boundaries between objects of normal use or consumption and collectibles, placing the latter in windows, shelves, albums, hanging them on walls, and so on. Sometimes the message, "This is a collectible object, don't use that as a trivial object," is clear for the collector and the external observer. But at other times the frames tend to fade even in the mind of the collector, and not only in the case of junk or tat objects. For example it is possible to see people playing like children with their old and precious pinball machines, jukeboxes or magic lanterns and sharing their enjoyment with friends.

Robert Opie, Jona and Peter's son, has founded his own Museum of Advertising and Packaging in Gloucester. But he lives in a crammed home with "boxes, tins, bottles, trade signs" and other stuff. Actually, as Robert said to Elsner and Cardinal in an interview: "There's nothing which I can not find a reason for saving." His archaeology is the every day life around him, his hunting ground. Boundaries fade in the words of Peter about his usual supermarket experience, "Have I actually got anything for dinner? I have only bought that I didn't want. Then I'll think: Can I find something I want to eat that is also useful for the collection?" (1994, pp. 25-48).

The oneiric house of Fiorella and Sandro Perolini, in their seventies, is one more example of happy and blurred cohabitation between collectors, things and semiophores. They are two well-known—among the specialists—Roman collectors of folkloric objects, of global craftsmanship, of memorabilia. The house encloses, like an eclectic postmodern Wunderkammer, an upper unbound set of citations of collections, real or virtual. There are, obviously, well consolidated thematic nuclei (cribs, pottery whistles, toys, devotional iconography, etc.), which are an important patrimony of folkloric goods sometimes used by the National Museum of Arts and Popular Traditions of Rome for its temporary exhibitions. However, the amused and poetic curiosity of Fiorella and Sandro, unchanged through the years, doesn't preclude the eclectic ephemera, generating multiple and overlapping citations of possible drifts of new collecting, even with a few items. The shelves can contain a couple of quirky salt and pepper shakers, souvenirs, carillons, pens, odd glasses, erotic or grim jokes, ending with grasping clues of the latest fashion of McDonald's Happy Meal.

As for Robert Opie, probably one of the keys for understanding something of their particular construct is the personal ability to reframe ironically the objects (Danet & Katriel, 1989, p. 261) and, to some extent, to reframe ironically themselves. They often play with the visitor the intentional role of tricksters, acting for her/him small and wonderful performances. As for a touch of a magician, toys animate, carillons play, radios sing, skulls walk. Visiting their home is like the Alice's journey into the world behind the mirror, where Fiorella and Sandro play, in turn, the amusing role of the Mad Hatter or of the Cheshire Cat.

They live in their collections that form a whole with the domestic symbols and commodities. Their relationship with the objects is one additional evidence—pace Pomian—of "the fact there is no firm user/collector dichotomy" (Akin, 1995, p. 103). In their long careers as collectors, they have changed and sometimes ignored the rules of their unique, idiosyncratic play. It has been enough to have a great deal of intellectual curiosity motivated by pleasure, irony, taste, sensitivity and capacity of discovering humor and the flare of poetry in the more humble and neglected objects.

REFERENCES

Akin, M. (1996). Passionate possession. The formation of private collections. In W. D. Kingery (Ed.), *Learning from things, method and theory of material culture studies* (pp. 102-128). London: Smithsonian Institution Press.

Alexander, E. P. (1996). *Museums in motion. An introduction to the history and functions of museums.* London: AltaMira Press.

Appadurai, A. (Ed.). (1986). *The social life of things: Commodities in cultural perspective.* Cambridge: Cambridge University Press.

Avedon, E. M. & Sutton-Smith, B. (Eds.). (1971). *The study of games.* New York: Wiley & Sons.

Balzac, H. de (1994). *Le cousin Pons: Les parents pouvres.* Paris: Pocket.

Bateson, G. (1972). *Steps to an ecology of mind.* New York: Ballantine.

Bateson, G. (1978). Play and paradigm. In M.A. Salter (Ed.), *Play: Anthropological perspectives* (pp. 7-16). New York: Leisure Press.

Baudrillard, J. (1968). *Le Système des objects.* Paris: Gallimard.

Belk, R. W. (1994). Collectors and collecting. In Pearce, S. M. (Ed.), *Interpreting objects and collections* (pp. 317-326). London: Routledge.

Benjamin, W. (1982). *Das passagenwerk, Gesammelte Schriften.* Band V, 1, Aufzeichnungen und Materialen (1928-1940). Frankfurt: Suhrkamp Verlag.

Bloom, J. (1997). *Baseball card collecting and popular culture.* Minneapolis, MN: University of Minnesota Press.

Borofsky, R. (Ed.). (1994). *Assessing cultural anthropology.* New York: McGraw-Hill.

Bourdieu, P. (1984). *Distinction: A social critique of the judgment of taste.* Cambridge, MA: Harvard University Press.

Bowman, J. R. (1978). The organization of spontaneous adult social play. In M. A. Salter (Ed.), *Play: Anthropological perspectives* (pp. 239-250). New York: Leisure Press.

Caillois, R. (1961). *Man, play and games.* Glencoe: IL: Free Press.

Chatwin, B. (1988). *Utz.* New York and London: Penguin.

Clifford, J. & Marcus, G. E. (Eds.). (1986). *Writing culture, the poetics and politics of ethnography.* Berkeley, CA: University of California Press.

Clifford, J. (1988). *The predicament of culture.* Cambridge, MA: Harvard University Press.

Danet, B. & Katriel, T. (1989). No two alike: Play and aesthetics in collecting. *Play & Culture 2* (3), 253-77.

de Saussure, F. (1973). *Course in general linguistics.* London: Fontana.

Diderot, D. & D'Alembert, J.l. (1751-1780). *Encyclopédie ou Dictionaire raisonné des sciences des arts et des métiers.* Nouvelle impression en facsimilé de la première édition de 1751-1780. Stuttgart: Badlannstatt (1966).

Elsner, J. & Cardinal, R. (Eds.). (1994). *The cultures of collecting.* Cambridge, MA: Harvard University Press.

Geertz, C. (1995). *After the fact: Two countries, four decades, one anthropologist.* Cambridge, MA: Harvard University Press.

Goffman, E. (1975). *Frame analysis. An essay on the organization of experience.* Harmondsworth, UK: Penguin Books.

Goldoni, C. (1977). La Famiglia dell'antiquario o sia la suocera e la nuora. In *Commedie di Carlo Goldoni: A cura di Nicola Mangini,* Vol. 1. Torino: Utet.

Greenblatt, S. (1991). Resonance and wonder. In I. Karp & S. D. Lavine (Eds.), *Exhibiting cultures, the poetics and politics of museum display* (pp.42-56). London: Smithsonian Institution Press.

Guggenheim, P. (1997). *Confession of an art addict*. Hopewell, NJ: Ecco Edition.

Harris, J. C. (1981). Beyond Huizinga: Relationships between play and culture. In A. Cheska (Ed.), *Play as context* (pp. 26-36). New York: Leisure Press.

Henricks, T. (1999). Play as ascending meanings: Implications of a general model of play. *Play and Culture Studies, 2,* 257-278.

Hughes, L. (1983), Beyond the rules of the game: Why are Rooie rules nice? In F. Manning (Ed.). *The world of play* (pp.188-199). New York: Leisure Press.

Huizinga, J. (1955). *Homo ludens: A study of the play element in culture.* Boston: Beacon.

Impey, O. & MacGregor, A. (Eds.). (1985). *The origins of museums: The cabinet of curiosities in sixteenth and seventeenth century Europe.* Oxford: Oxford University Press.

Kingery, W. D. (Ed.). (1996). *Learning from things, method and theory of material culture studies.* Washington and London: Smithsonian Institution Press.

Lugli, A. (1983). *Naturalia et Mirabilia. Il collezionismo enciclopedico nelle Wunderkammern d'Europa.* Milano: Mazzotta.

Lyotard, J. F. (1979). *La condition postmoderne.* Paris: Les Edition de Minuit.

MacGregor, A. (1985). The cabinet of curiosities in sixteenth-and seventeenth-century Britain. In O. Impey & A. MacGregor (Eds.), *The origins of museums: The cabinet of curiosities in sixteenth and seventeenth century Europe* (pp. 147-158). Oxford: Clarendon Press.

Mead, G. H. (1934). *Mind, self and society.* Chicago, IL: University of Chicago Press.

Mullen, C. (1994). 'The people's show'. In S. Pearce (Ed.), *Interpreting objects and collections* (pp.287-290). London: Routledge.

Pearce, S. M. (1993). *Museums, objects and collections.* Washington: Smithsonian Institution Press.

Pearce, S. M. (1994). Collecting reconsidered. In S. M. Pearce (Ed.), *Interpreting objects and collections* (pp. 193-204). London: Routledge.

Pearce, S. M. (1998). *Collecting in contemporary practice.* London: Sage.

Pearce, S. M. (Ed.). (1994). *Interpreting objects and collections.* London: Routledge.

Pertinaz, D. (1996). *Homo collector, museos insólitos, collecciónes extravagantes, collecionistas exéntricos.* Barcelona: Pandora.

Pomian, K. (1990). *Collectors and curiosities: Paris and Venice, 1500-1800.* Cambridge: Polity Press.

Rosenberg, S. (1970). *The come as you are masquerade party: William James Sidis, the streetcar named paradise lost.* Englewood Cliffs, NJ: Prentice Hall.

Sack, A. L. (1977). Sport: Play or work. In P. Stevens Jr. (Ed.), *Studies in the anthropology of play: Papers in memory of B. Allan Tindall* (pp. 186-195). New York: Leisure Press.

Sassower, R. (1995). *Cultural collisions, postmodern technoscience.* New York and London: Routledge.

Sontag, S. (1993). *The volcano lover: A romance.* New York: Doubleday.

Stewart, S. (1984). *On longing: Narratives of the miniature, the gigantic, the souvenir, the collection.* Baltimore: John Hopkins University Press.

Stocking, G. W. Jr. (Ed.). (1985). Objects and others, essays on museums and material culture. *History of anthropology, Vol. 3.* Wisconsin: The University of Wisconsin Press.

Sutton-Smith, B. (1997). *The ambiguity of play.* Cambridge, MA: Harvard University Press.

Sutton-Smith, B. (1999). Evolving a consilience of play definitions: Playfully. In Reifel S. (Ed.), *Play and Culture Studies, Vol. 2* (pp. 239-256). Stamford, CT: Ablex.

von Neumann, J. & Morgenstern, O. (1944). *Theory of games and economic behavior.* Princeton: Princeton University Press.

Windsor, J. (1994). Identity parades. In J. Elsner & R. Cardinal (Eds.), *The cultures of collecting* (pp. 49-67). Cambridge, MA: Harvard University Press.

Wittgenstein, L. (1953). *Philosophical investigations.* Oxford: Basil Blackwell.

Part V

Conclusion: Playfulness and Tricksters

Whether or not the prior collecting history is or is not fully valid, what is clear is that what is idealized there is play behavior of a most playful conceptual kind. What we appear to have done in this book, therefore, is to pass play forward from the less controllable evolutionary forms to those that seem structured systemically by the various different kinds of cultures, including our own schools, and then have moved on to the increasing variability, individuality and imaginativeness, with or without toys, of the more recent modern kinds of play. It is not surprising in these terms that playfulness has become increasingly idealized in modern play theory in today's individually oriented as compared with yesterday's collectively oriented cultures, (or independent versus interdependent cultures as framed by Greenfield and Cocking, 1994). Not surprising either that much modern play research work has been done on the playfulness of children. What it shows is that those who are playful are more spontaneous in all respects (social, physical, and intellectual), as well as humorous and joyful, and are in general in a better personal, social and academic status than those who do not score so highly on these attributes (Barnett, 1998; Singer, 1999).

Having made these claims for the moderns over the ancient, however, we are immediately faced with multiple problems. To begin with, there is a sense in which much of the variety of what one sees as an exercise of individuality in the modern sense can also be called a kind of conspicuous consumption, though perhaps more of an experiential kind than the economic kind discussed by Veblen (1899) in his theory of the leisure class. There is much consumer oriented wishful thinking about much modern play when it is defined in some recent research as flow, or in the zone, or self actualization or peak experience, and as such might be seen by a modern skeptic as having as more to do with the collectivity of consumer economics than individualism. Again, even the children's playfulness, as positive as it might seem in current studies, may well be concealing more negative aspects of play that are also a part of normal peer play activities. This is illustrated in Linda Hughes' (1993) wonderful revelation of the co-operative kind of appearances she found in the game of four square played in a Quaker school, where despite these appearances some of the players were masking underlying Mafia like arrangements favoring their own clique. And in a more general way such deceptions, and even cruelties, are diversely illustrated by many playground studies (Sutton-Smith, 1990). Again with adults, McMahon has shown that there is a variety of ways in which subordinate groups such as women position themselves through their group play relationships to subvert male control in and out of the play world (McMahon, 1993). These behaviors include dark play during which men who are playing on one level do not realize that there is another second level of play going on among the women of which they are the targets. To return to "playfulness," in most of these studies there is some occasional mention also of these outstanding players having more provocative characteristics such as teasing and mischief, but these kinds of elements are not generally much heeded given the rationalistic categorization of modern childhood in orthodox developmental psychology (Sutton-Smith, 1997, chapter 9). One gets a better grasp of the modern child's playful expression in such practical advocacy as in Gary Krane's *Simple Fun for Busy People* (1998) which is full of examples of more provocative, hilarious and absurd types of playfulness. But the best source at present for the ultimate character of the possibilities of playfulness is probably the legendary trickster literature which is about the analysis and character of various tricksters throughout history. This analysis includes the legendary descriptions of such characters as Till Eulenspiegel, Prometheus, Wakdjunkaga, Coyote, Brer Rabbit, Pulcinello, Harlequin, Jester, Hermes, Krishna, Windigo, Bugs Bunny, and Maui to mention just a few (Christen, 1998; Hyde, 1998). Their chief characteristics are their endlessly variable cunning intelligences as well as their grotesque bodily manifestations. They are said to practice irony, ambiguity, ambivalence, satire, parody, paradox, persiflage, and all and sundry kinds of rule breaking inversions. But at the same time they are irrepressible and inexhaustibly optimistic in the face of incredible temerity (Koepping, 1985). Most of all they are boundary crossers defiling right and wrong, sacred and profane, clean and dirty, male and female, young and old, living and dead. "In every case trickster will cross the line and confuse the distinction" (Hyde, 1998, p.7). What we achieve from all of this laughter in the face of anxiety is apparently victory over fear as well as a self-affirmation despite the banality, chaos and mortality of everyday life. What accompanies such trickster play is said to be a sense of emotional restoration and sense of emotional plenitude. All of which is a part apparently of the mobile freedom that this kind of play indulges. Unfortunately, however, this is the kind of analysis of the legendary past now being lauded by some modern theorists, as an illustration of how randomly modern play might actually be. But that means if there is any historical verity to these legendary accounts, the ancient tricksters at least, if not everyone, had as mobile an imagination as we are claiming for our modern selves by these analyses. What the trickster concept does for our playfulness is unite it with the playfulness concept throughout the history of human beings. But that means that the kind of history from the collective to the individualistic that we have been tracing is not the entire truth. As far back as we can trace the play behavior of humans we have illustrations of such ludic trickery. And given the statements by Gould and Fagen earlier about randomness and non adaptive preferences, as well as what we know about animal play and as well as what Edwards claimed was universal in her six cultures, we have no reason not to believe that here is a basic potential novelty of playfulness at birth in all human creatures. It has been noted that even religious occasions were playful in many early cultures. And it has been observed that in some quite ancient and some still existing foraging cultures in Africa, Australia and America there has been observed an abundance of play variety amongst the young (Marshall, 1976; Salter, 1969) All of which means that what has been mainly illustrated by the cross cultural and school articles in this work are examples of the adult systemic cultural overlays which often confine children's play in ways that decrease its spontaneous variability even though it is often still quite remarkable.

INTRODUCTION AND INTERSTITIAL REFERENCES

Abt, C. C. (1981). *Serious games.* New York: Viking.

Barnett, L. A. (1998). The adaptive powers of being playful. In M. C. Duncan, G. Chick & A. Aycock. (Eds.), *Play and culture studies, 1* (pp. 97-120) Greenwich, CT.; Ablex.

Boyer, P. (2001). *Religion explained.* New York: Basic Books.

Berlyne, D. E. (1960). *Conflict, arousal and curiosity.* NewYork: McGraw-Hill.

Briggs, J. L. (1998). *Inuit morality play.* New Haven, CT: Yale University Press.

Bruner, J. S, Jolly, A. & Sylva, K. (1976). *Play.* New York: Basic Books.

Burghardt, G. M. (1998). The evolutionary origins of play revisited: Lesson from turtles. In M. Bekoff & J. A. Byers (Eds.), *Animal play: evolutionary, comparative and ecological perspectives* (pp. 1-28). New York: Cambridge University Press.

Burghardt, G. M. (1999). Conceptions of play and the evolution of animal minds. *Evolution and Cognition, 5,* 115-123. ·

Burghardt, G. M. (2001). Attributes and neural substrates. In E. Bass (Ed.), *Handbook of behavioral neural biology,* 13:327-366. New York: Plenum Press.

Carson, J., Burks, V. & Parke, R. D. (1993). Parent-child physical play determinants and consequences. In K. MacDonald (Ed.), *Parent-child play, descriptions and implications* (pp. 197-220). Albany: SUNY.

Carse, J. (1986). *Finite and infinite games.* New York: Free Press.

Chick, G. (1998). Games in culture revisited: a replication and extension of Roberts, Arth & Bush. *Cross-cultural research, 32*: 2: 185-206.

Christen, K. A. (1998). *Clowns and tricksters.* Denver, CO: ABC-Clio.

Cross G. (1997). *Kids stuff: The changing world of American childhood.* Cambridge, MA: Harvard University Press.

D'Agnese. J. (2002). An embarrassment of chimpanzees. *Discover, 23*:42-49.

Damasio, A. R. (1994). *Descartes error: Emotion, reason and the human brain.* New York: Putnam.

Damasio, A. R. (1999). *The feeling of what happens.* New York: Harcourt.

DeLoache, J. & Gottlieb, A. (2000). *A world of babies.* NY: Cambridge University Press.

Diamond. J. (1992). *Guns, germs and steel.* New York: Norton.

Dorst, J. (1999). Which came first, the chicken device or the textual egg? Documentary film and the limits of the hybrid metaphor. *Journal of American Folklore, 112*:445 (Summer): 268-281.

Drewal, M. T. (1991). *Yoruba ritual: performers, play, agency.* Bloomington, IN: Indiana University Press,

Fagen, R. (1981). *Animal play behavior.* New York: Oxford.

Fagen, R. (1995). Animal play, games and angels: Biology and brain. In A. D. Pellegrini (Ed.), *The future of play theory* (pp. 23-44). Albany, NY: SUNY Press.

Falassi, A. (1967). *Time out of time: essays on the festival.* Albuquerque, NM: University of New Mexico Press.

Fernandez, J. W. (1986). *Persuasions and performances: The play* of *tropes in culture.* Bloomington, IN: Indiana University Press.

Gabler, N. (1998). *Life the movie.* New York: Knopf.

Goodman, F. D. (1992). *Ecstasy, ritual and alternate reality.* Bloomington, IN: Indiana University Press.

Goncu, A. (Ed.) (1999). *Children's engagement in the world.* New York: Cambridge University Press.

Gould, S. J. (1996). *Full house.* New York: Harmony Books.

Greenfield, P. M. & Cocking, R. R. (1994). *Cross cultural roots of minority child development.* New Jersey: Erlbaum.

Guttman, A. (1978). *From ritual to record: the nature of modern sports.* NY: Columbia Press.

Herron, R. & Sutton-Smith, B. (Eds,) (1971). *Child's play.* New York: Wiley.

Houseman, M. (2001). Playful power and ludic spaces; studies in games of life. *Focal: European journal of anthropology, 37*:39-48.

Howard, D. (1977). *Dorothy's world: Childhood in Sabine Bottom 1902-1910.* New York: Prentice-Hall.

Hughes, L. (1993). Children's games and gaming. In B. Sutton-Smith, J. Mechling, T. Johnson & F. McMahon, (Eds.), *Children's folklore: A source book* (pp. 93-120). Logan, UT: Utah State University Press.

Huizinga, J. (1955/1938). *Homo Ludens: A study of the play element in culture.* London: Routledge & Kegan Paul.

Hutcheon, L. (1985). *A theory of parody.* New York: Methuen.

Hutt, C. (1971). Exploration and play in children. In R.E. Herron & B. Sutton-Smith (Eds.), *Child's play.* New York: Wiley.

Hyde, L. (1998). *Trickster makes the world.* New York: Farrar, Strauss & Giroux.

Koepping, K. P. (1985). Absurdity and hidden truth, cunning intelligence and grotesque body images and manifestations of the trickster. *History of Religion, 24*: 3:191-213.

Krane, G. (1998). Krane, G. (1998). *Simple fun for busy people: 333 Free Ways to Enjoy Your Loved Ones More in the Time You Have.* Berkeley, CA: Conari Press.

Lancy, D. F. (2001). Cultural constraints on children's play. *Play and Culture, 4*:53-62.

Marshall, L. (1976). *The! Kung of NyaeNyae.* Cambridge, MA: Harvard University Press.

McClary, A. (1997). *Toys with nine lives: A social history of American toys.* North Haven, CT: Linnet Books.

McMahon, F. (1993). The worst piece of "tale": Flaunted 'hidden transcripts' in women's play. *Play Theory & Research, 1*: 4: 251-258.

McMahon, F. (2000). The aesthetics of play in reunified Germany's carnival. *Journal of American Folklore, 113*:450 (Fall) 378-390.

Messner, M. A. (1990). When bodies are weapons: Masculinity and violence in sport. *International Review For Sociology of Sport, 25*: 3: 203.

Miracle, A. (1974). Some functions of Aymara games and play. In P. Stevens (Ed.), *Studies in the anthropology of play* (pp.98-104). West Point, NY: Leisure Press.

Moore, R. (1986). *Childhood's domain: play and place in child development.* Berkeley, CA: MIG Communications.

Nelson, M. B. (1994). *The stronger women get, the more men love football.* New York: Harcourt Brace & Co.

Nuwer, H. (1999). *Wrongs of passage: Fraternities, sororities, hazing and binge drinking.* Bloomington, IN: Indiana University Press.

Piaget, J. (1951). *Play dreams and imitation in childhood.* NewYork: Norton.

Power. T. G. (2000). Play and exploration in children and animals. New Jersey: Erlbaum.

Roberts, J. M., Arth, M. J. & Bush, R. R, (1959). Games in culture. *American Anthropologist, 61*,597-605.

Roberts, J. M. & Sutton-Smith, B. (1962). Child training and game involvement. *Ethnology, 1*, 166-185.

Rojek, C. (1995). *Decentring leisure.* London: Sage.

Roskos, K. A. & Christie, J. F. (2000). *Play and literacy in early childhood.* New Jersey: Erlbaum.

Shore, A. N. (1994). *Affect regulation and the origin of the self.* New Jersey: Erlbaum.

Singer, D. G. (1999). Imagination, play and television: Factors in a child's development. In J. A. Singer & P. Salovey, (Eds.), *At play in the fields of consciousness* (pp. 303-336). New Jersey: Erlbaum.

Spariosu, M. (1989). *Dionysus reborn.* Ithaca, N.Y: Cornell, University Press.

Sutton-Smith, B. (1959). *The games of New Zealand children.* Berkeley, CA: University of California Press.

Sutton-Smith, B. (1990). The school playground as a festival. *Children's Environment Quarterly, 7*: 2,3-7.

Sutton-Smith, B. (1992). The role of toys in the instigation of creativity. *Creativity Research Journal, 5*,1:3-11.

Sutton-Smith, B. (1997). *The ambiguity of play.* Cambridge, MA: Harvard University Press.

Sutton-Smith, B. (1998). *Review of Emory conference: The culture of toys.* TASP Newsletter, Vol. 23:1.

Sutton-Smith, B. (2001). Emotional breaches in play and narrative. In A. Goncu & E. L. Klein. *Children in play, story and school* (pp. 161-176). New York: Guilford Press.

Sutton-Smith, B. (2002). Recapitulation redressed. In Conceptual, social-cognitive, and contextual issues in the field of play. *Play & Culture, 4*, pp. 3-20.

Sutton-Smith, B., Mechling, J., Johnson, T. W. & McMahon, F. (Eds.) (1999). *Children's folklore.* Logan, UT: Utah State University Press.

Tuan, Y. F. (1998). *Escapism.* Baltimore, MD: John Hopkins University Press.
Veblen, T. (1899). *The theory of conspicuous consumption.* New York: Macmillan.
Wegener-Sphoring, G. (1989). War toys and aggressive games. *Play and Culture, 2,* 1, 35-47.
Wrangham, R. & Peterson D. (1996). *Demonic males.* New York: Houghton-Mifflin.

About the Contributors

KRISTIE ANDREWS [MSEd., Monmouth University, Reading, 2002] is a third grade teacher at the Mill Lake School, Monroe Public Schools, Monroe, New Jersey, USA.

ANITA C. BUNDY [ScD., Boston University, Therapeutic Studies, 1987] is Professor in the Department of Occupational Therapy at Colorado State University in Ft. Collins, Colorado, USA.

PAOLA dE SANCTIS RICCIARDONE [Demo-ethno-anthropological disciplines, State Universities National Selection, Italy 1995] is Associate Professor of Cultural Anthropology and Material Culture in the Department of Archeology and History of the Arts at University of Studies of Calabria (UNICAL), Cosenza, Italy.

CAROLYN POPE EDWARDS [Ed.D., Harvard University, Human Development, 1974] is Professor of Psychology and Family and Consumer Sciences at the University of Nebraska--Lincoln, in Lincoln, Nebraska, USA.

ROBERT FAGEN [Ph.D., Harvard University, Biomathematics 1974] is Associate Professor, University of Alaska Fairbanks in Juneau, Alaska, USA.

JEFFERY GLINER [Ph.D., Bowling Green State University, Experimental Psychology, 1971] is Professor in the Department of Occupational Therapy at Colorado State University in Ft. Collins, Colorado, USA.

ROBYN M. HOLMES [Ph.D., Rutgers University, Cultural Anthropology, 1988] is Associate Professor in the Dept. of Psychology, Monmouth University in West Long Branch, New Jersey, USA.

ANNICA LÖFDAHL [Ph.D., Karlstad University, Pedagogy 2002]. is employed at Department of Education, member of the research group for development and learning, Karlstad University, Sweden.

DONALD E. LYTLE [Ph.D., United States International University, Human Behavior and Leadership, 1984] is a Professor in the Department of Kinesiology, California State University, Chico, California, USA.

CHERYL R. MEAKINS [M.S., Colorado State University, Occupational Therapy, 2002] is a Contract Occupational Therapist with Foothills Gateway, Fort Collins, Colorado, USA.

KALYANI D . MENON [Ph.D., Syracuse University, Department of Anthropology, 2002] is a part-time instructor in the Department of Anthropology, Syracuse University in Syracuse, New York, U.S.A.

FELICIA McMAHON [Ph.D., University of Pennsylvania, Folklore/Folklife Studies, 1992] is Research Associate in the Dept. of Anthropology, Program on the Analysis and Resolution of Conflicts, Syracuse University in Syracuse, New York, USA.

PEGGY O'NEILL-WAGNER [LATG, Undergraduate and Graduate work in Psychology, University of Wisconsin,1975] Work dedicated to the well-being of nonhuman primates since 1971. Joined the National Institutes of Child Health and Human Development staff as a Research Psychologist in 1984. Specializes in early development and enrichment of captive primates at the NIH Animal Center, Poolesville, Maryland, USA

THOMAS L. REED [PhD., University of South Carolina, Early Childhood Education, 1996) is an assistant professor of early childhood education in the School of Education at the University of South Carolina ~Spartanburg, Spartanburg, SC, USA.

LYNN ROMEO [EdD., Rutgers University, Curriculum, 1984] is Associate Professor, Chair, and Graduate Program Director in the Department of Educational Leadership & Special Education, Monmouth University, West Long Branch, New Jersey, USA.

JEAN-PIERRE ROSSIE (Ph.D., State University of Ghent, Belgium, African History and Philology, 1973) is Member of the Stockholm International Toy Research Center, Royal Institute of Technology, Stockholm, Sweden.

BRIAN SUTTON-SMITH [Ph.D., University of New Zealand, Developmental Psychology, 1954] is Emeritus Professor, University of Pennsylvania, Departments of Education and Folklore, USA.

ARNE TRAGETON [M.Ed., University of Trondheim, Norway, Pedagogy, 1977] is Associate Professor at Stord/Haugesund University, Norway.

Recent Titles in *Play & Culture Studies* Series